MW01528401

Atlantic Sentinel

Atlantic Sentinel

Newfoundland's Role
in
Transatlantic Cable Communications

by D. R. Tarrant

Flanker Press Ltd.
St. John's, Newfoundland
1999

Printed in Canada by Robinson-Blackmore Ltd.
St. John's, Newfoundland, Canada

Published in Canada by
Flanker Press Ltd.
P O Box 2522, Station C, St. John's
Newfoundland, Canada   A1C 6K1

Tel/Fax (709) 739-4477
email: flanker@roadrunner.nf.net
Website: www.flankerpress.com

Canadian Cataloguing in Publication Data

Tarrant, Donald R., 1946-

    Atlantic sentinel

ISBN 1-894463-00-5

1. Cables, Submarine -- Atlantic Ocean -- History.
2. Telegraph -- Newfoundland -- History. 3. Telegraph
stations -- Newfoundland -- Heart's Content -- History.
4. Telegraph stations -- Newfoundland -- Bay Roberts --
History. I. Title.

TK5611.T37 1999   384.1'09718   C99-950150-X

Donald Tarrant, P. Eng., is a professional engineer who has more than thirty years experience in the communications industry. He presently resides in St. John's, Newfoundland Canada.

CONTENTS

PROLOGUE

ONE
THE ELECTRIC TELEGRAPH..1

TWO
FREDERICK NEWTON GISBORNE 7

THREE
CYRUS FIELD ENTERS THE SCENE 19

FOUR
THE ATLANTIC CONQUEST BEGINS
     – THE 1857 ATTEMPT..27

FIVE
SUCCESS AND DISAPPOINTMENT
     – THE 1858 ATLANTIC CABLE 39

SIX
OTHER PROPOSALS 49

SEVEN
SO CLOSE BUT YET SO FAR
     – THE 1865 ATLANTIC CABLE 53

EIGHT
TRIUMPH AT LAST
     – THE 1866 ATLANTIC CABLE 61

NINE
EXPANSION OF TRANSATLANTIC TELEGRAPHY............71

TEN
ANGLO-AMERICAN/WESTERN UNION
    – HEART'S CONTENT..75

ELEVEN
WESTERN UNION TELEGRAPH COMPANY
    – BAY ROBERTS...97

TWELVE
DIRECT UNITED STATES CABLE COMPANY
    – HARBOUR GRACE..107

THIRTEEN
COMMERCIAL CABLE COMPANY
    – ST. JOHN'S...115

FOURTEEN
FROM TELEGRAPH TO TELEPHONE CABLES.................129

FIFTEEN
CANADIAN OVERSEAS TELECOMMUNICATION
    CORPORATION (COTC)..133

SIXTEEN
END OF AN ERA..139

ENDNOTES...141

BIBLIOGRAPHY..145

APPENDIX – MAJOR SUBMARINE CABLES
    REFERENCED..147

INDEX..153

PROLOGUE

Newfoundland's geographic position as North America's nearest landfall to Europe has given it a prominent role in the fields of transportation and communications. The province has been the point of departure or destination for countless adventurers, including Alcock and Brown, who flew the Atlantic nonstop in 1919, and Charles Lindbergh, who marked St. John's as his last sight of land during his solo transatlantic flight to Paris in 1927. Others have departed Newfoundland using the natural forces of wind and current to attempt to fly the Atlantic by balloon, or row across it in the tiniest of rowboats. More foolhardy individuals have attempted to pedal across it in semi-submersibles or drift across it in a barrel.

Many communities have gained prominence because of Newfoundland's proximity to Europe. During both World Wars, St. John's was an important port for transatlantic shipping. Gander, which is used as a stop for transatlantic flights, and Botwood, which served as a refuelling stop for the flying boats in the 1930s, have also benefited because of their locations. The same can be said of the transatlantic cable towns of Heart's Content, Bay Roberts and Harbour Grace. These and other communities have played important roles in the province's history. Newfoundland's importance in the history of transatlantic transportation and communications must therefore not be underestimated. Since its discovery by John Cabot in 1497, Newfoundland has been the guardian or sentinel of the North American continent. To reach North America, Newfoundland had to be approached first. Therefore, ATLANTIC SENTINEL, the title of this book.

Newfoundland's location, less than 2000 miles from Europe, made it the logical location for the landing of the first transatlantic telegraph cable. This remarkable event captured the imagination of millions of people around the world and is the focus of the pages that follow. Newfoundland's place in communications history is confirmed by pioneers such as Frederick Gisborne and Cyrus Field. However, Newfoundland's transatlantic cable stations also played important roles in overseas communications, and here too is an exciting story to be told.

This short history covers the period from the mid 1800s to the late 1990s. During this era, Newfoundland's first telegraph system (1851) was installed, the first transatlantic cable (1858) was laid, the first transatlantic radio message (1901) was received, the first off-island telephone message (1939) was made, the first transatlantic telephone cable (1956) went into service, and the first off-island digital optical fibre system (1991) was installed. The history of communications in Newfoundland has been fascinating, and there have been many significant firsts.

In the mid 1800s, Newfoundland's population was barely 100,000. Its only industry was fishing and the only regular communication in the colony was provided by the postal service. There were no roads to speak of, and the only link between the hundreds of communities and the outside world was the sea. Coastal boats plied the coast of Newfoundland and Labrador and connected the area to Montreal, the Canadian maritime provinces and the eastern coast of the United States. The electric telegraph was in its infancy, and in the mid-1800s finally reached Newfoundland. It is with the electric telegraph that the story of Newfoundland's role in transatlantic communications begins.

The last of the cable stations in Newfoundland closed almost thirty-five years ago, and few former cable company employees still survive. In addition, many were transferred to other parts of the world where their employers still had overseas telegraph operations. I have attempted, wherever information was available, to name the staff members of the cable companies and thereby give recognition to those who contributed so significantly to Newfoundland's role in transatlantic telegraph communications.

I would like to acknowledge with thanks the invaluable assistance, information and advice I received in preparing this volume. I would like to thank the staffs of the Public Archives of Newfoundland and Labrador (PANL), the Newfoundland

Collection of the A. C. Hunter Library, the Memorial University Centre for Newfoundland Studies, and the Heart's Content cable museum. Among the individuals I would like to thank are Robert Balsom, Ron Brown, Sandra Clarke, Gertrude Gerstle, Jack Hambling, Claude Hobbs, Gus Kerwin, Basil Martin, Willis P. Martin, Raymond Norman, Neil Legge, Edgar Parke, Bruce Perry, Ron Stapleton, William B. Trickett, W. Hubert Trickett, Richard Wade, Alton Whelan, and Audrey (Mrs. Ted) Withers.

Thanks are also given for permission to reproduce photographs. Credit is given with each photograph, and those without acknowledgement are part of my personal collection.

D. R. Tarrant,
St. John's, 1999

One

THE ELECTRIC TELEGRAPH

Before the invention of electric telegraphy in the mid 1800s, communications between distant points relied on messages delivered by personal courier, ships, mail, newspapers, carrier pigeons, and various other means of transportation. More primitive ways of communicating messages over great distances used drums, flashing lights, flags, and smoke signals to indicate events. A cannon blast, for instance, could announce the arrival of a ship; however it could not communicate the name of the ship, its time of arrival or any other detailed information.

In the late 1700s, semaphore systems came into use. Semaphore used highly visible moveable arms on high posts which were manipulated in different configurations to represent letters of the alphabet. Stations were installed on hills or high buildings, separated from one another as far as the eye could see. This distance was normally several miles. Semaphore was capable of transmitting complete messages and was the first practical system that could transmit words and sentences with any reliability.

Operators at semaphore stations received messages from the previous station and transmitted them to the next, so there was obviously lots of room for errors and misinterpretations. Some systems were more than 1000 miles long. Two notable European systems extended between Paris and Toulon, a distance of approximately 475 miles, and between Prussia and Leningrad (via Warsaw), a distance of 1200 miles. The latter required 1300 operators to send a message from one end to the other.

One of the earliest semaphore systems in British North America was built in 1799 between Halifax and Annapolis, Nova Scotia. This was constructed by the British army to enable them to communicate with their troops between these locations.[1]

Sending messages by semaphore was labourious and time consuming, not only because of the number of times each message had to be repeated, but also because each message had to be verified before it was relayed to the next station. Semaphore was primarily a terrestrial communications medium and could not be used over wide bodies of water beyond the visual range of the operators. It was used chiefly for government and military communications and was not generally available for public messages. Yet, despite its shortcomings, semaphore was employed for long distance communications until electric telegraphy came into use.

EARLY TELEGRAPH SYSTEMS

In the early 1800s, there were dozens of different electric telegraph systems, either in operation or in the proposal stage. One was the "chemical" telegraph, built by Sommerring in Munich in 1809. This system employed wires and tanks of water for each letter of the alphabet so that when an electrical connection was sent over a particular wire, bubbles would rise from the tank affected, thus representing a particular letter. Another electric telegraph system, built in 1816 by Sir Francis Ronald at Hammersmith, England, employed static electricity. In this system, static charges were sent over separate wires to deflect lightweight balls at the far end of the circuit to expose letters of the alphabet. Both of these telegraph systems, along with a number of others, worked after a fashion, but were totally impractical for widespread use.[2]

Practical electronic telecommunications began in 1819 with Hans Christian Oersted (1777-1851). Oersted was a Danish physicist and one of the first to experiment with electromagnetism. In 1819, he noted that a magnetic needle would deflect if it was placed next to a wire carrying electricity. The needle would deflect to the right or the left, depending on the direction in which the electric current was flowing through the wire. This was a major discovery because it demonstrated a direct physical manifestation of the heretofore theoretical sciences of electric fields and electromagnetism. Oersted did not recognize a practical use for this discovery; however, shortly after his findings were published, other scientists and inventors made use of his work and went on to develop useful applications. One of the more important was the electric motor. Another was the electric telegraph.

To produce early electric telegraphs, inventors built devices to change the direction of electric currents through wires in pre-determined sequences. This created various deflections of one or more magnetic needles at the end of the circuit. Particular sequences of deflections represented individual letters of the alphabet.

The most common telegraph system during the nineteenth century Europe was designed by two Englishmen, William Cooke and Charles Wheatstone. It was introduced in 1835 and required the sender at the transmitting end to make certain electrical connections, each of which represented a different letter of the alphabet. Each connection caused magnetic needles at the receiving end to deflect in a certain sequence which enabled the receiver to determine the letter being transmitted. The operation was slow, and required as many as five wires between stations. Despite these drawbacks, Cooke and Wheatstone's telegraph was used by the British railways for several years, and was later improved so that only two wires and two needles were required. Other telegraph systems also evolved in Europe; however, none proved to be as practical as one invented on the other side of the Atlantic.

SAMUEL MORSE

North America's first practical electric telegraph was developed by Samuel Finley Breese Morse (1791-1872). Morse was born in Charlestown, Massachusetts, and spent his first forty years pursuing an interest in visual arts. In the 1830s, he was a professor of painting and sculpture at the City University of New York. Morse was an accomplished painter, and some of his works in historical art included portraits of Eli Whitney, Noah Webster, the Marquis de LaFayette, and President James Monroe.[3] Besides fine arts, he also had an interest in science, and held patents on gadgets as diverse as hydraulic pumps and marble carving machines. In 1832, while returning from Europe, he heard about the properties of electro-magnetism and immediately took an interest in the subject. This interest in science overtook his career as an artist and prompted his experimentation in this new area. His work in this field eventually led him to telegraphy and other areas of communications.

Morse recognized that electromagnetism could be applied to an improved system of telegraphy and developed a device to demonstrate this. His early instrument employed an electromagnet to deflect a pencil to make marks on a strip of paper. The paper was drawn beneath a pencil at a continuous speed by a spring-wound clockwork mechanism. By making and breaking an electric circuit at different intervals at the transmitting end, the signal received at the other end would cause the pencil to leave marks on the paper strip. By breaking the circuit in certain sequences, patterns of these marks represented letters of the alphabet.

Early telegraphers deciphered the message by analyzing the paper strips, but it was not long before they began use the clicking sound of the electromagnet to decipher them. This unique sequence of dots and dashes became known as Morse Code. Later models of the telegraph took advantage of this audio property by employing the electromagnet to attract a piece of metal on a spring to create a clicking sound. This audio system was more practical, and the paper recorder fell into disuse.

Samuel F. B. Morse (1791-1872) (Courtesy of the Provincial Archives of Newfoundland and Labrador (PANL))

Morse demonstrated his first working telegraph in 1836. However, it was not until 20 June 1840 that his patent application was approved. It took another four years before the electromagnetic telegraph caught the public's eye.

Morse set up the first practical telegraph circuit between Washington, D.C. and Baltimore, Maryland, with the help of a $30,000 grant from the U.S. Congress. He demonstrated it on 24 May 1844 from the Capitol in Washington, D.C. to an audience of

congressmen and other dignitaries. The first message was a passage from Numbers 23:23 in the Bible – "What hath God wrought?" It is not known if the question was answered.

It did not take long for entrepreneurs to realize the economic potential of Morse telegraphy. Shortly after the Capitol demonstration, a company was set up in May 1845 to build a line to New York. This organization, named the Magnetic Telegraph Company, quickly expanded throughout the United States.[5]

Morse became renowned as a scientist and inventor throughout the United States and the world. He became involved with Newfoundland's communications history through his association with Cyrus Field and the first transatlantic telegraph cable, which will be discussed in more detail later.

THE EXPANSION OF ELECTRIC TELEGRAPHY TO EASTERN CANADA

Electric telegraphy expanded quickly throughout North America. In the mid 1800s, telegraph lines were installed between the major cities in the United States, and telegraphy soon became an integral part of the communications infrastructure. Telegraph lines were extended to Buffalo, New York in 1846, and shortly afterwards to Canada.

The first telegraph in Canada was demonstrated in Toronto, Ontario on 24 July 1846. Before the end of the year, the first telegraph line in Canada was installed between Hamilton and Toronto by the Toronto, Hamilton & Niagara Electro-Magnetic Telegraph Company. The first message over the system was transmitted on 19 December 1846. On 14 January 1847, the line was connected to Albany and Buffalo, New York, via a cable which crossed the Niagara River from Ontario to Lewiston on the American side.[6] In the 1840s, the cost of sending a message was expensive, up to three shillings sixpence for the first ten words.[7]

The Toronto, Hamilton & Niagara Electro-Magnetic Telegraph Company was soon purchased by the Montreal Telegraph Company (Montreal Telegraph). Montreal Telegraph set up a line between Quebec City, Montreal and Toronto in 1847, and later made connections to Michigan and the New England states. Montreal Telegraph underwent considerable expansion; by 1875, it had more than 20,000 miles of wire, 12,044 miles of pole line, 1507 offices and 2330 employees.[8]

Orrin Wood, a protege of Samuel Morse, was the superintendent of Montreal Telegraph, while Frederick Gisborne was the company's chief operator in Quebec City. Gisborne was to make a tremendous contribution to Newfoundland's communications history, as discussed in the next chapter.

1849 saw the expansion of electric telegraphs to eastern Canada. In that year, the first telegraph line in the region was set up by Lawson R. Darrow. It ran between Saint John, New Brunswick and Calais, Maine, a distance of 80 miles.[9] Later that year, the telegraph system was also extended to Halifax.

Two

FREDERICK NEWTON GISBORNE

The story of Newfoundland's first telegraph begins in 1851 with the arrival in St. John's of Frederick Newton Gisborne (1824-1892). Gisborne was born in Broughton, Lancashire, England on 8 March 1824, and as a youth was taught mathematics and engineering by the town vicar. Prior to coming to Canada, he travelled widely, and visited Australia, New Zealand, Tahiti, Guatemala and Mexico. In Tahiti, Gisborne and his uncle experimented in cultivating trees which produced gutta-percha, a latex-like substance which among other things was used as an electrical insulator. In 1845, his travels took him to Canada where he worked on a farm in St. Eustache, Quebec. He became interested in communications, and spent most of his spare hours studying electricity and telegraphy. In 1846 he quit

Frederick Gisborne (1824-1892) (Courtesy of PANL)

farming and accepted a position with Montreal Telegraph, where he trained as an operator under Orrin Wood. With the experience gained with Montreal Telegraph, he became an expert in electric telegraphy and a year later organized the British North American

Electric Telegraph Association, whose objective was to extend telegraphic communications from Montreal to the east coast of Canada.[10] While head of this association, he supervised the construction of a 112-mile long telegraph line between Quebec City and Rivière du Loup.

Gisborne headed east with the view of convincing the New Brunswick government to build a system connecting New Brunswick with the Quebec telegraph network. Unfortunately, the government was more interested in establishing telegraph connections to Maine in the United States, and rejected his suggestion. In 1849, he found himself in Halifax, Nova Scotia, where he became General Superintendent of the Nova Scotia Telegraph Company. By that time, the telegraph system had already been extended from Calais, Maine to Saint John and Amherst. His first priority therefore, was to construct a telegraph line between Amherst and Halifax. With the completion of this link, Halifax was connected into the North American telegraph system. This was of immense importance to the Halifax business community, particularly the newspaper industry, as the inclusion of Halifax in the North American telegraph system enabled European news arriving by ship to be quickly telegraphed to North American newspapers and printed in their next editions. This provided news to the American market considerably faster than via ships arriving in New York.

The newspapers' thirst for news from Europe greatly impressed Gisborne and he set his mind on extending the telegraph system to St. John's, Newfoundland, North America's most easterly city. Gisborne believed that connecting St. John's to the North American telegraph system would make it a major terminus for transatlantic news. Transatlantic steamers would stop at St. John's to drop off news, which would immediately be relayed by telegraph to North American newspapers. News from Europe could therefore be provided to North American newspapers 24 to 48 hours earlier than from transatlantic steamers arriving at Halifax or New York.

At the time, St. John's was not a regular port of call for transatlantic steamers, and the success of Gisborne's plan would have an obvious economic impact on the city, as well as on the owners of the telegraph line. There were hundreds of North American newspapers vying for news information, and the business was extremely competitive. The first newspaper to receive news from Europe would have a great advantage over its rivals, and would have been willing to pay handsomely for that privilege.

BISHOP MULLOCK

Among the first to support the notion of connecting Newfoundland to the North American telegraph system was J. T. Mullock, the Roman Catholic Bishop of Newfoundland. Mullock was involved in many political issues of the day, and had strong views on the role of steamships and railways, in addition to communications. In 1850 he was on his boat, visiting parishioners in the small remote communities along the southwest coast of Newfoundland. While sailing around Cape Ray, he noted the short distance between Cape Breton Island and Newfoundland, and wondered why the two islands could not be connected by telegraph. He saw that it might be possible to traverse the Cabot Strait with an underwater cable, linking Newfoundland to the North American telegraph system. This would provide vastly improved communications between Newfoundland and the rest of North America, yielding great economic benefits to the country. In a letter to the *St. John's Courier* on 8 November 1850, he expressed his views:

I regret to find that in every plan for Transatlantic Communication Halifax is always mentioned and the natural capabilities of Newfoundland entirely overlooked. This has been deeply impressed on my mind by the communications I read in your paper of Saturday last, regarding telegraphic communication between England and Ireland, in which it is said that the nearest telegraphic station on the American side is Halifax, twenty-one hundred and fifty-five miles from the west of Ireland. Now would it not be well to call the attention of England and America to the extraordinary capabilities of St. John's as the nearest telegraphic point? It is an Atlantic port, lying, I may say, in the track of the ocean steamers, and by establishing it as the American telegraphic station, news could be communicated to the whole American continent forty-eight hours, at least, sooner than by any other route. But how will this be accomplished? Just look at the map of Newfoundland and Cape Breton. From St. John's to Cape Ray there is no difficulty in establishing a line passing near Holy-Rood along the neck of land connecting Trinity and Conception Bays, and thence in a direction due west to the Cape. You have then about forty-one to forty-five miles of sea to St. Paul's Island, with deep sound-ings of one hundred fathoms, so that the electric cable will be perfectly safe from icebergs. Thence to Cape North, in Cape Breton, is little more than 12 miles. Thus it is not only practica-ble to bring America two days nearer to Europe by this route, but should the telegraphic communication between England and Ireland, 62 miles, be realized, it presents not the least difficulty. Of course we in Newfoundland will have nothing to do with the erection, working, and maintenance of the telegraph; but I sup-pose our Government will give every facility to the company, either English or American, who will undertake it, as it will be an incalculable advantage to this country. I hope the day is not far distant when St. John's will be the first link in the electric chain which will unite the Old World and the New.

Mullock clearly had the foresight and imagination to visu-alize a telegraph cable between Newfoundland and the mainland. However, he was a churchman, not an engineer or businessman. It would take the entrepreneurial spirit and knowhow of individuals such as Frederick Gisborne and Cyrus Field to bring Mullock's vision to reality.

GISBORNE'S PETITION TO THE NEWFOUNDLAND GOVERNMENT

While in Halifax in early 1851, Gisborne started planning to extend the telegraph system to St. John's. Among his first steps were two petitions to the Newfoundland government. The first was for a grant to build a telegraph line between St. John's on the east coast of the island and Cape Ray on the southwest coast, which would connect by submarine cable to the Nova Scotia tele-graph system. The petition was endorsed by Samuel G. Archibald, a prominent St. John's citizen and an acquaintance of Gisborne. It was introduced in the Newfoundland legislature by Philip F. Little, a member of the House of Assembly for St. John's, who would later become Newfoundland's first prime minister. On Gisborne's petition, Little stated in the House of Assembly:

> The Petitioners offer to construct a telegraph line; pro-vided an act is passed incorporating them as the "Newfoundland Telegraph Company" and that they receive a grant to assist them in effecting their surveys and in opening a road or path along such telegraph lines and the exclusive right to such tele-

graph for a term of forty years, and of such ungranted land along such line, as may be necessary to farm; and similar privileges for a line to Trepassey.

Gisborne's second petition was for a grant to construct a telegraph line between St. John's and Carbonear, as well as a line to Trepassey. Gisborne believed that this system would be of great benefit to the business community as well as being profitable to its operators. On introducing the petition in the legislature, Edmund Hanrahan, the member from Conception Bay, described part of the project:

> ... to construct a line of Electric telegraph between St. John's and Carbonear, which they submitted would be a great accommodation to the mercantile community, and remunerative to its constructors, taking the distance by land at sixty miles, they are prepared to erect a line on the most approved principles for £2000, being the rate of £30 per mile; with stations at St. John's, Bay Roberts, Harbour Grace and Carbonear, and put them into efficient working order.

After being advised that the government was receptive to his petitions, Gisborne in a letter dated 11 March 1851 wrote to Archibald:

> Should the Government see fit to grant me the necessary funds for the St. John's and Carbonear telegraph line, I shall leave for the former place immediately not only to make the necessary arrangements for erecting that line so soon as the frost is out of the ground, but also for the purpose of agitating for the extension of the line from St. John's to Cape Race and Trepassey. The whole could be erected at a cost of about £12,000, and the income would be very considerable as I shall presently prove to you. First, however, let me explain to you a practical plan of operation, until the submarine part of the line were laid, that is, even supposing the steamers do not call at St. John's or Trepassey. Off Cape Race I should place the yacht *Wanderer*, belonging to the Associated Press of New York, which has already been offered to me for that purpose. She

would intercept most of the steamers bound west and receive the news from them in signal kegs prepared for that purpose. The yacht would then run for Trepassey, where I should have a station, thence the news would be telegraphed instantly via St. John's to Cape Ray, where it would be written out on slips and flown by-carrier-pigeons (of which we have fifty trained ones on hand) to Cape North in the island of Cape Breton, a distance of only about sixty miles, and thence telegraphed to New York; so that we could put the leading items of the news through from Trepassey to New York, or New Orleans, within three hours of daylight.

For this service alone, even should we succeed in intercepting a steamer but once in four voyages, and should succeed in putting the news through in three hours instead of four or five days, in advance of the steamer's arrival at New York, the Associated Press would gladly pay the Newfoundland line alone from £2000 to £3000 per annum. Last year the New York press paid the lines between Halifax and New York upwards of £6000 for the news, although received but from one to sixty hours in advance of the arrival of the steamers. In fact the news will command any price; so that without any private business, the line would yield a large profit...

But this is not all – So soon as you can practically show that such an immense advantage arises from Newfoundland as a news Station, you must inevitably secure for St. John's or Trepassey the attention of both England and America, and the speedy result will be the creation of a Packet Station at either of these places. All that is required is speedy and energetic measures on the part of the representatives of the people, to secure to the country that inestimable advantage. So pray be up and stirring, and the next year you may be in a fair way of securing a packet station for transatlantic steamers.

Gisborne resigned as chief officer of the Nova Scotia Telegraph Company late in the summer of 1851 and sailed to Newfoundland to begin implementing his plans. He appeared before the legislature in St. John's and presented his proposal for a telegraph line from that city to Cape Ray with future plans to extend it via a submarine cable to Cape Breton Island. The government was impressed with Gisborne's presentation and gave him permission to proceed, as well as £500 to survey a route across the island.[11] However, before tackling the route to Cape Ray, Gisborne began work on the St. John's – Carbonear line.

ST. JOHN'S AND CARBONEAR ELECTRIC TELEGRAPH COMPANY

By September 1851, the St. John's and Carbonear Electric Telegraph Company had begun construction of a telegraph line along the highway between the two towns. Once the line reached Brigus, more than half way to Carbonear, Gisborne let the construction crew carry on, while he branched west to survey the route toward Cape Ray and to make preparations for his major project – a proposed line across the island.

The cross-island survey covered almost 400 miles of harsh woods and barrens over mainly uncharted territory, and took several months to complete. The expedition was plagued by extremely bad weather, the worst in living memory at the time. During the survey, one of Gisborne's party of six men died and others suffered from cold weather and a shortage of supplies. Despite those problems, the survey was completed by the end of the year.

Meanwhile, the St. John's – Carbonear line was progressing favourably, and work was completed by December. However, there were problems in finding telegraph operators, and since they had to be trained, the line did not go into service until 6 March 1852. A news item on its opening was printed five days later in the 11 March edition of the *Newfoundlander:*

> The Electric Telegraph between St. John's and Conception Bay was put in operation for the first time on last Saturday, and has transmitted several messages from Brigus and Harbour Grace each day of this week. Yesterday particularly, the Telegraph Office was the scene of general attraction throughout the day.

As part of the inauguration of the Carbonear line, Gisborne gave a public talk to the Mechanics Institute in St. John's. The lecture was held on 6 March 1852 at the Old Factory, and was attended by government officials and other dignitaries. The highlight of the evening was a demonstration of the system, and for the occasion, a telegraph line was set up between the Old Factory and the telegraph office at the Commercial Building on Duckworth Street. The telegraph office was already connected into the system linking it to Harbour Grace, Carbonear and Brigus.

During the talk, Gisborne demonstrated the telegraph line by exchanging messages between St. John's and the outlying communities.

The St. John's – Carbonear telegraph system was a commercial success. News from the Conception Bay area reached St. John's with little delay, and newspapers prefaced news from the area with the headline "per Electric Telegraph." The line was heavily used by businesses and residents; however, after a few weeks it went out of service because of vandalism caused by boys throwing rocks at the insulators. The St. John's and Carbonear Electric Telegraph Company requested compensation for damages from the Newfoundland government and was awarded a payment of £100 to make repairs and put the line back in operation.

THE MAINLAND LINE

With the St. John's – Carbonear telegraph system in place, Gisborne turned his attention to the line to Cape Breton Island. Having already obtained approval from the Newfoundland government for an overland telegraph line to Cape Ray and a submarine cable across the Cabot Strait, he approached the government of Nova Scotia for their approval.

Gisborne wrote the Nova Scotian government requesting permission to land a cable at Cape North, Cape Breton Island, and to extend it overland to Sydney. To his surprise, the government rejected his proposal. It was obvious the Nova Scotia government was concerned that Halifax would lose a good part of the news forwarding telegraph business to St. John's if the cable from Newfoundland was completed, and it denied his request to land a submarine cable in Nova Scotia.

Undaunted by this setback, Gisborne looked at other alternatives for connecting Newfoundland to the mainland. He decided that a submarine cable from Cape Ray to Prince Edward Island and then on to New Brunswick would be the next best choice. Although this required a greater length of underwater cable, the overall distance from Newfoundland to New York City would be shorter than the proposed route via Cape Breton Island.

NEWFOUNDLAND ELECTRIC TELEGRAPH COMPANY

The Newfoundland legislature passed an act incorporating the Newfoundland Electric Telegraph Company (Newfoundland

Electric) in the spring of 1852, giving it the right to construct a telegraph line between St. John's and Cape Ray. The company was also granted the right to construct branch lines to Trepassey and other locations, as well as the exclusive rights to all telegraphy in Newfoundland for a thirty-year period.

Newfoundland Electric's authorized capital was £100,000, consisting of 1000 shares of £100 each. The company was organized in New York, and headquartered at 19 Trinity Buildings. New York businessmen Horace B. Tebbets and Darius B. Holbrook were its major shareholders. Tebbets was involved in the steamship business and was planning a New York to Ireland route. The idea of diverting his ships to St. John's to deliver European news was obviously a factor in his interest in the company.

In New York, Gisborne learned that John and Jacob Brett had installed an underwater telegraph cable between England and France. John Watkins Brett, at forty-five years old, was a retired antiques dealer who became interested in telegraphy. His brother Jacob was an engineer who shared the same interest, and they set up a company for a submarine cable across the English channel. The Bretts obtained a two-year concession from the British and French governments for the telegraph connection. C. J. Wollaston, one of the directors of Brett's English Channel Submarine Telegraph Company, was the engineer in charge of the project.[12]

The cross-channel cable was manufactured by the Gutta Percha Company and consisted simply of a copper conductor covered in gutta-percha, a rubber-like substance obtained from the tree of the same name, grown mainly in Malaysia. At warm temperatures, the substance is soft and pliable, whereas at colder temperatures, it hardens. One of its more useful properties is that it is an excellent insulator for wires and cables.

Since submarine telegraphy was in its infancy, few engineers, including Brett and Wollaston, understood that there was a great difference in the electrical characteristics of underwater cables and those suspended from poles. The Bretts' telegraph line would have been fine for above-water use, but proved not to be suitable for the task at hand.

The Bretts laid their cable from Dover to Cape Grisnez, France in 1850 with only a few days to spare before their concession expired. Because the Bretts were rushed for time, engineering and installation mistakes occurred, and as a result, the cable was not successful. After carrying only a few messages, which for the most part were incoherent, the cable completely failed the day after its installation, when it was accidentally pulled up by a French fisherman. Although the project was a failure, it proved that an underwater link would work, and that a better-designed cable would be required.

In September 1851, a more successful cross-channel cable was installed with the involvement of T. R. Crampton, an English engineer, who not only designed the system, but also put up 50% of the cost of the project. The cable was made of solid copper wire insulated with gutta-percha reinforced with a protective iron sheath.

Encouraged by the success of the English channel cable, Gisborne sailed to England in 1852 and met with John and Jacob Brett to discuss details of the cross-channel cable. He purchased a supply of underwater cable from Newall & Company which he intended to be part of a system to connect Newfoundland with Nova Scotia via St. Paul's Island. However, as noted earlier, this plan went nowhere because the Nova Scotia government denied him landing rights. Gisborne instead used the cable to extend the telegraph system across the Northumberland Strait between Cape Tormentine, New Brunswick and Carleton Head, Prince Edward Island. The fifteen-mile long cable was installed by the *Ellen Gisborne* in November 1852 and was North America's first practical submarine telegraph cable installation.

Gisborne and John Brett kept in touch by mail and even discussed proposals regarding a transatlantic cable. Brett proposed setting up a company to place a telegraph line between Ireland and Newfoundland, and estimated that the project would cost about £750,000. He informed Gisborne that he could raise half the amount in England if Gisborne could raise the other half in North America. As obtaining this amount of money was well beyond Gisborne's capability, he dropped the idea and set his mind on completing the line to Cape Ray. Any thoughts he might have had for a transatlantic cable were put aside.

Gisborne began construction of the St. John's – Cape Ray line in the summer of 1853. The project employed 350 men and started at Brigus, where it connected into the St. John's – Carbonear system. Gisborne initially buried the cable in order to protect it from the weather. However it did not take him long to realize the impracticality of digging in the rocky Newfoundland terrain, and he wisely proceeded with a pole line instead.

After only 40 miles of construction from St. Johns, Gisborne's New York backers failed to honour his bills, and the project came to a sudden halt. Newfoundland Electric became insolvent and Gisborne went bankrupt.

In 1854, Gisborne went to New York to see Horace Tebbets to solicit funding to revive the company. Tebbets, for unknown reasons, was no longer interested in the telegraph line, and Gisborne had to look elsewhere. He even wrote to John Brett inquiring if he was interested in purchasing the remains of Newfoundland Electric. Brett would later state that he was interested in investing in the project; however before he could respond, Gisborne looked elsewhere for financing.

At Astor House in New York, fate intervened when Gisborne met Matthew Field, a civil engineer, who introduced him to his brother Cyrus. With Gisborne's introduction to Cyrus Field, communications in Newfoundland took a turn, and a new era in telegraph communications began.

Discussing the transatlantic cable. Left to Right: Peter Cooper; David D. Field, Chandler White, Marshall Roberts, Samuel F. B. Morse, D. Huntington, Moses Taylor, Cyrus W. Field, Wilson G. Hunt (Courtesy of Heart's Content museum)

Three

CYRUS FIELD ENTERS THE SCENE

Cyrus W. Field (1819-1892) was an American merchant, who at the age of thirty-three had made a fortune in the paper business. Having left his business ventures behind, he was in 1854 looking forward to a comfortable retirement. He had just returned from a six-month vacation in South America when his brother Matthew told him about Frederick Gisborne's thoughts on extending the North American telegraph system to Newfoundland. Field was so interested he invited Gisborne to his home at 1 Lexington Avenue in Gramercy Park, New York City, where the two discussed Gisborne's ideas regarding extending the telegraph system across the Cabot Strait to Cape Ray and then on to St. John's. Late in the evening, after Gisborne had left, Field was contemplating his discussion with his guest, and went to his study to take a closer look at his globe to check on Newfoundland's location. The extension of the North American telegraph system to St. John's made sense to him. However, his attention was caught by the short distance between Newfoundland and Europe. How surprised he was to discover that Newfoundland was 1000 miles closer to Europe than New York

Cyrus Field (1819-1892)
(Courtesy of PANL)

was! The notion of extending telegraphic communications across the Cabot Strait would have been by itself a huge undertaking, but Field envisioned a far greater challenge. Why stop at Newfoundland? Why not extend the system all the way to Europe?

Field speculated as to whether a telegraph cable could be placed between Newfoundland and England along the bottom of the Atlantic Ocean. The economic benefits to the owner of such a cable would obviously be considerable. As a businessman, Field immediately seized the significance of Gisborne's plan to extend the telegraph system to Newfoundland, and envisioned the commercial potential of extending it even farther to Europe. He gave up all ideas of retirement and began the pursuit of a transatlantic cable, a notion which would be his obsession for the next twelve years.

THE NEW YORK, NEWFOUNDLAND AND LONDON TELEGRAPH COMPANY

Field's first priority was to set up a company to implement the transatlantic cable. It was named the New York, Newfoundland and London Telegraph Company (New York Telegraph), the principals of which were the "five immortals," who besides Field, included the New York capitalists Peter Cooper (president), Moses Taylor (treasurer) and Marshall O. Roberts and Chandler White (directors). Wilson G. Hunt became a director the following year. The company was organized on 10 March, 1854, when the principals of New York Telegraph bought from Newfoundland Electric the "telegraph lines, wires, posts, insula-

Discussing the setup of the New York, Newfoundland and London Telegraph Company (Ocean Telegraphy – Twenty fifth anniversary)

tors, cables, appurtenances, apparatus, land privileges and bonuses and all other property of this Company on the island of Newfoundland and on Prince Edward Island and in the province of New Brunswick and in the waters between New Brunswick and Prince Edward Island . . ."[13] The assets were sold to Peter Cooper, Moses Taylor, Cyrus W. Field, Marshall Roberts and Chandler White for $40,000. Newfoundland Electric's steam yacht *Victoria* was also sold to the same group for $5,000. The five original shareholders initially invested $500,000 in the company but eventually put additional millions of dollars into the enterprise.

David Dudley Field, Cyrus's brother, was counsel for the firm. Gisborne was appointed Chief Engineer; however, this was a token gesture on Field's part, and Gisborne quickly faded from any position of importance.

Cyrus Field made the first of many visits to Newfoundland in March 1854. He was accompanied by his brother David, as well as Frederick Gisborne and Chandler White. They left New York on 14 March, via Boston, and sailed to Halifax where they picked up the steamer *Merlin* for St. John's. [14] There they met Edward. M. Archibald, Newfoundland's Attorney General, who introduced them to Ker Baillie Hamilton, the governor of Newfoundland from 1852 to 1855. Field returned to New York after a three-day stay but the others continued discussions with government officials in St. John's. The group remaining eventually met with the legislative council and laid out their plan to build a telegraph system to connect Newfoundland with the mainland.

An area of extensive discussion involved the charter of the bankrupt Newfoundland Electric Telegraph Company. This was important to New York Telegraph because it needed a monopoly on landing telegraph cables in Newfoundland to make the transatlantic project feasible. It eventually obtained the government's consent and agreed to take over the charter, rights and franchise of Newfoundland Electric and to pay off its outstanding wages and debts. In return, the government agreed to guarantee a 5% interest on New York Telegraph's £250,000 bond for twenty years, waive the duties on all wires and cables imported by the company and grant it £5000 for the construction of a road along the telegraph line. The government also offered fifty square miles of land upon completion of the telegraph system to Cape Breton, and an additional fifty square miles of land upon completion of the transatlantic line to Ireland. Most importantly, the company was also to receive a fifty-year monopoly on telegraphs in the country, including the landing of cables from North America and Europe.

The formation of New York Telegraph was approved by the Newfoundland legislature on 15 April 1854. In 1867, it approved an increase in the company's capital to $6,000,000 to allow for the construction of additional land lines as well as a submarine cable from Placentia, Newfoundland to mainland North America.

New York Telegraph's monopoly was one of the earliest in the world granted to a communications company. The Newfoundland government's action caused a great furor in London, and the British government advised its Colonial and Provincial legislatures that in the future Her Majesty would be advised not to give her ratification to the creation of similar monopolies. Regardless of the British government's stand, Field succeeded where Gisborne had failed, and negotiated an arrangement with Nova Scotia, as well as with Prince Edward Island, the government of Canada, and the state of Maine.[15] The monopoly was a perennial issue in the late 1800s and early 1900s, when it was used to prevent other cable companies from locating in Newfoundland.

It is interesting to note that this monopoly was also used in the early 1900s to prevent Marconi from setting up transatlantic wireless communications facilities in Newfoundland. Instead, Marconi set up his North American base at Glace Bay, Nova Scotia.

THE START OF THE OVERLAND LINE
With the charter secure, Chandler White and Matthew Field sailed to St. John's in the spring of 1854 to organize the activities of New York Telegraph.[16] White handled the administrative duties, while Field provided the engineering and construction supervision. They quickly arranged for the construction of an overland cable across Newfoundland over almost 400 miles of wilderness. The project involved the construction of a pole line and an eight-foot wide path along the route. Many bridges also had to be built to traverse the rivers the line had to cross. The route followed the south coast of the island, which allowed the work crews to be provisioned by steamer. Supplies were delivered by

the company's ship *Victoria* to various locations along the coast and were carried inland to the site of the cable construction. The line was to be built in eleven sections, each employing its own work crew:

Section	Length (miles)
St. John's to Piper's Hole	122
Piper's Hole to Long Harbour	34
Long Harbour to Conne River	20
Conne River to North Bay	30
North Bay to White Bear Bay	57
White Bear Bay to Grandy's Brook	25
Grandy's Brook to Couteau Bay	23
Couteau Bay to Garia Bay	35
Garia Bay to Burnt Islands	20
Burnt Islands to Port aux Basques	23
Port aux Basques to Cape Ray	10

The project employed 600 men and, although expected to finish in 1855, was not completed until October 1856. A submarine cable still had to be placed across the 90-mile wide Cabot Strait to connect Newfoundland with the North American telegraph system. This project was scheduled to be completed in August 1855 with a cable consisting of three conductors manufactured by the English company Kuper & Co. (later to become Glass, Elliot & Co.).

THE CABOT STRAIT CABLE

Field, the consummate promoter, wanted the world to know about the proposed line between Newfoundland and the North American mainland and he seized the opportunity to publicize the occasion. In anticipation of the great event, he organized a group of journalists, scientists, company officials and other invited guests, and on 7 August departed New York aboard the steamer *James Adger* for the Cabot Strait to witness the installation. Also on board was Samuel Morse, who was Field's adviser on technical matters.

En route from New York, Field wined and dined his guests. At Halifax some disembarked and returned home via Saint

John, New Brunswick. The others continued on the *James Adger* to Port-aux-Basques to witness the installation. Upon arrival, Field learned that Samuel Canning, the project's supervisor, had already arrived by steamer from England; however the barque *Sarah L. Bryant*, which was delivering the cable from England, was a few days behind schedule. While waiting for the *Bryant*, Field and his entourage sailed to St. John's and met with public officials. Within a few days, they were back in Port-aux-Basques to find that the *Sarah L. Bryant* had arrived and was ready to start the project.

Since the *Bryant* was a sailing vessel, it was not sufficiently manoeuvrable to lay the cable on its own, and had to be towed by the *James Adger*. The project started smoothly; however, when the vessels were about forty miles from Cape Ray, a vicious storm came up. The closest point of land was St. Paul's Island, off the coast of Cape Breton, and despite the high winds, the ships tried to land the cable there. Unfortunately the storm was too strong and the *Bryant's* captain decided he had no option but to cut the cable in order to save the barque from sinking. Both ships headed to the nearest harbour for protection. With not enough cable for another attempt, the project was abandoned.

"Terminus at Cape Ray, Newfoundland," from the Illustrated London News, 20 Oct 1855

Cyrus Field, Samuel Morse, the journalists and the rest of the entourage were presented with an unexpected and embarrassing failure, not the historic occasion that they were expecting. The *James Adger* and its disappointed passengers steamed back to New York.

Field remained confident nonetheless that a cable could be successfully laid across the Cabot Strait between Newfoundland and Nova Scotia, and realized that this connection had to be made for his greater scheme of a transatlantic cable to succeed. He therefore made arrangements with Kuper & Company in England to manufacture a new cable. Since this company was anxious to prove their underwater cable was commercially viable, it agreed to provide the cable at its own risk.[17]

Having learned from the experience of the *Sarah L. Bryant* that a sailing vessel was not suitable for laying submarine cable, Field arranged for a steamship to next perform the task. In 1856, he procured the steamship *Propontis*, and geared it up with new stranded copper cable. Under the command of Captain Goodwin, the ship successfully completed the installation across the Cabot Strait to Nova Scotia without incident. The first phase of Field's grand plan was finally completed, and for the first time, Newfoundland was connected by telegraph to the North American mainland.

The telegraph system to St. John's had cost more than a million dollars, of which Field personally paid twenty percent. It was far more difficult than expected and took more than two years to finish. Some time later, Field reflected on the difficulty of building the Newfoundland telegraph system in the following comments:

> To extend a line from Nova Scotia across Newfoundland to St. John's, thence across the ocean, was a very pretty plan on paper. There was New York, and there was St. John's, only about 1200 miles apart. It was easy to draw a line from one point to the other, making no account of the forests and mountains and swamps and rivers and gulfs that lay in the way. None of us had ever seen the country or had any idea of the obstacles to be overcome. We thought we could build the line in a few months. It took us two and a half years. Yet we never asked help outside our own little circle. Indeed, I fear we should not have got it if we had, for few had any faith in the scheme. Every dollar came out of our own pockets..... Our only support outside was in the liberal Charter and steady friendship of the Newfoundland Government, at that time.

Despite the setbacks in building the line, telegraphic communication from St. John's to the rest of North America was now a reality. The first commercial message over the system took place on 1 October 1856, when J.& W. Pitts, a dealer in provisions, groceries and rum in St. John's, received a message from Baddeck, Cape Breton Island.[18]

ALEXANDER M. MACKAY

However, all was not well with the Newfoundland overland line and service was frequently disrupted. By December 1856 the Cape Ray line had gone dead and nobody seemed to know how to keep it properly maintained. In order to save his investment, Field desperately needed technical assistance, so he contacted his business friends for advice. In discussions with two of his associates, James Eddy and D. H. Craig, the founder of the *New York Associated Free Press*, Field asked for advice on who could be found to repair and operate the Newfoundland cable system. They recommended a young twenty-two year old Nova Scotian named Alexander M. Mackay.

Mackay at the time was superintendent of the Nova Scotia Telegraph Company. He was contacted by Field and convinced to resign and take employment with New York Telegraph. He moved to Newfoundland in 1857 and found the telegraph line completely inoperable. Mackay and a work crew walked the route between Cape Ray and St. John's and made repairs, where necessary rebuilding the line. The cost of this work amounted to $90,000. By June, a break in the Cabot Strait was also repaired and telegraphic communication to Cape Breton was reestablished.

Alexander McLennan Mackay (1834-1905) (Courtesy of PANL)

Now that Newfoundland was again linked to North America, the scene was set for Field to start to bridge the gap across the North Atlantic.

Four

THE ATLANTIC CONQUEST BEGINS – THE 1857 ATTEMPT

In the early years of electronic communications, connecting Europe and North America by telegraph was pondered by many visionaries of the day. Samuel Morse had considered the idea in the 1840s, and in 1843 stated in a letter to the United States Secretary of the Treasury that "a telegraphic communication on the electric-magnetic plan, might with certainty be established across the Atlantic Ocean. Startling as this may seem now, I am confident the time will come when this project will be realised."[19] Bishop Mullock had proposed the idea in 1850, and Gisborne and John Brett discussed the concept several years later. Connecting North America and Europe with a telegraph line was obviously not a unique idea and others, (some of whom are discussed later in this volume), had had similar thoughts.

It was through the efforts of Cyrus Field that a transatlantic cable began to appear possible. Although Field – as much as he sometimes appeared to think to the contrary – did not originate the idea of extending a submarine cable crossing the Atlantic Ocean, it is only proper to credit him for having recognized the commercial implications of a transatlantic cable and having the skill and determination to bring this idea to fruition.

THE TECHNICAL EVALUATION

Having realized the economic benefits of a telegraph connection between North America and Europe after his meeting with Gisborne in early 1854, Field began looking into the business details of laying the cable. He quickly recognized the engineering problems of such a venture and set his mind to finding answers to his concerns. What about the Atlantic currents? Could a cable long enough and strong enough be manufactured? What kind of ship would be needed to lay it? Could a telegraphic signal be sent across the width of the Atlantic Ocean? All were difficult questions needing answers and he wrote letters to prominent scientists seeking their advice. One letter was to Samuel Morse, asking if it was possible to send a telegraphic signal over such a long distance. Another was to Lieutenant Matthew F. Maury, head of the National Observatory in Washington, requesting undersea survey information on a possible transatlantic route.

Morse was the most prominent telegraph scientist of the day and took a strong interest in the idea of a transatlantic telegraph cable. Within a few days of receiving Field's letter, the two men met in New York. From this time on, Morse was a close adviser to Field on technical matters relating to the transatlantic project. He told Field about his experiments with underwater cables and advised him that a transatlantic telegraph system was technically feasible. With this assurance, Field had the comfort he needed and proceeded to look into other aspects of the project.

In response to Field's inquiry, Lieutenant Maury responded that substantial information was already available from a survey conducted by Lieutenant O. H. Berryman in 1853. Berryman's survey included measurements of wind and current as well as soundings along the ocean floor taken at regular intervals between Newfoundland and Ireland. The survey concluded that the route would be deep and flat enough so that an underwater cable would not be hindered by icebergs or ships' anchors. In his response Lieutenant Maury stated:

> This line of deep-sea soundings seems to be decisive of the question as to the practicability of a Submarine Telegraph between the two continents, in so far as the bottom of the deep sea is concerned. From Newfoundland to Ireland, the distance between the nearest points is about sixteen hundred miles; and the bottom of the sea between the two places is a plateau, which seems to have been placed there especially for the purpose of holding the wires of a Submarine Telegraph, and of keeping them out of harm's way. It is neither too deep nor too shallow;

yet it is so deep that the wires, but once landed, will remain for-ever beyond the reach of vessels' anchors, icebergs, and drifts of any kind, and so shallow that the wires may be readily lodged upon the bottom. The depth of this plateau is quite regular, grad-ually increasing from the shores of Newfoundland to the depth of from fifteen hundred to two thousand fathoms as you approach the other side. The distance between Ireland and Cape St. Charles, or Cape St. Lewis, in Labrador, is somewhat less than the distance from any point of Ireland to the nearest point in Newfoundland. But whether it would be better to lead the wires from Newfoundland or Labrador is not now the question; nor do I pretend to consider the question as to the possibility of finding a time calm enough, the sea smooth enough, a wire long enough, a ship big enough, to lay a coil of wire 1,600 miles in length; though I have no fear but that the enterprise and inge-nuity of the age, whenever called on with these problems, will be ready with a satisfactory and practical solution of them.

I simply address myself at this time to the question in so far as the bottom of the sea is concerned, and so far as that the greatest practical difficulties will, I apprehend, be found after reaching soundings at either end of the line, and not in the deep sea.... Therefore, so far as the bottom of the deep sea between Newfoundland, or the North Cape, at the mouth of the St. Lawrence, and Ireland, is concerned, the practicability of a Submarine Telegraph across the Atlantic is proved.

With encouraging reports from Morse and Maury, Field was satisfied that a transatlantic cable was feasible and over the next several years made more than forty trips to England to pur-sue the project. On his first trip, he ordered a length of submarine cable for the link between Newfoundland and Nova Scotia. While in England he also discussed plans for a transatlantic cable with John Brett, who had earlier been involved with underwater cables across the English Channel. Brett was impressed with Field's plans and was later to invest capital in Field's transatlantic cable company. Field was also supported by Professor William Thompson of the University of Glasgow, who had also contem-plated the idea of a transatlantic telegraph cable. Thompson was the inventor of much of the telegraph equipment used in the early transatlantic cables, and later became Lord Kelvin, after whom the temperature scale is named.

THE ATLANTIC TELEGRAPH COMPANY

Field decided to set up a new company for the transatlantic cable project. In September 1856, the Atlantic Telegraph Company (Atlantic Telegraph) was formed for the purpose of installing a telegraph cable across the Atlantic Ocean to connect in Newfoundland with New York Telegraph's line to the mainland. Field served as first vice-president and Charles Bright, the well-known British telegraph engineer, was engineer-in-chief. The Board of Directors included Sir William Brown as chairman, C. M. Lampson, John Pender, John Brett, Dr. Whitehouse and Professor Thompson.[20]

The Newfoundland government gave its approval for the Atlantic Telegraph Company to consolidate with the New York Newfoundland and London Telegraph Company on 3 March 1857. Although Atlantic Telegraph assumed New York Telegraph's monopoly at this time, the complete merger of both companies did not occur until 1873.

Field toured Britain addressing those interested in supporting the bold idea of a transatlantic cable. The public looked upon this scheme with considerable scepticism and the project was viewed by many to be a foolhardy, impossible undertaking. Despite public cynicism, Atlantic Telegraph's issue of 350 shares of £1000 each was fully subscribed within a month, with Field taking eighty-eight shares and Brett twenty-two. Field had control of about one quarter of the stock, which he intended to sell to American investors. Once he returned to the United States, however, he found that Americans generally doubted a transatlantic cable would succeed, and he had difficulty finding takers. In fact he sold only twenty-seven shares after several months of effort and paid for the remaining shares with his own finances.[21] Undaunted, he pushed on to continue the organization of the project.

The British Government passed an Act guaranteeing Atlantic Telegraph £14,000 per year (4% of the £350,000 capital) until the company's net profits were £10,000 per year, after which it would pay £10,000 per year. The British government also agreed to provide ships to conduct further ocean surveys. In return the Atlantic Telegraph Company would provide free messages to the government up to the annual guaranteed amount and would give it priority on all messages.

Field attempted to persuade the American government to accept a similar deal. The U.S. Congress balked at providing funding for a non-American enterprise involving a cable between two British territories. After considering the advantages a telegraph system between New York and London would bring to the country, Congress acceded to Field's request.[22] Despite much debate and opposition, it ratified the proposal, but only by the slimmest of margins. Since both the British and American governments agreed to essentially the same deal, it was decided that priority on messages would be given each government on a first-come-first-served basis.

THE SUBMARINE CABLE

Field realized that the submarine cable would have to be extensively tested before the transatlantic project could go ahead. With the help of the English and Irish Magnetic Company, experiments were performed to determine if electric telegraphy would work over the 2000 miles or so of cable that would be needed to cross the Atlantic. The company conducted experiments during October 1856 with ten reels of cable, each measuring more than two hundred miles in length, connected together to simulate the distance across the ocean. These experiments were successful, proving that telegraphy would work over such a distance. Morse wrote the following letter to Field describing the results of the testing:

London, five o'clock a.m.,
October 3, 1856

My dear Sir:
As the electrician of the New York, Newfoundland, and London Telegraph Company, it is with the highest gratification that I have to apprise you of the result of our experiments of this morning upon a single continuous conductor of more than two thousand miles in extent, a distance you will perceive sufficient to cross the Atlantic Ocean, from Newfoundland to Ireland.
The admirable arrangements made at the Magnetic Telegraph Office in Old Broad street, for connecting ten subterranean gutta-percha insulated conductors, of over two hundred miles each, so as to give one continuous length of more than

two thousand miles during the hours of the night, when the telegraph is not commercially employed, furnished us the means of conclusively settling, by actual experiment, the question of the practicability of telegraphing through our proposed Atlantic cable.

This result had been thrown into some doubt by the discovery, more than two years since, of certain phenomena upon subterranean and submarine conductors, and had attracted the attention of electricians, particularly of that most eminent philosopher, Professor Faraday, and that clear-sighted investigator of electric phenomena, to wit, the perceptible retardation of the electric current, threatened to perplex our operations, and required careful investigation before we could pronounce with certainty the commercial practicability of the Ocean Telegraph.

I am most happy to inform you that, as a crowning result of a long series of experimental investigation and inductive reasoning upon this subject, the experiments under the direction of Dr. Whitehouse and Mr. Bright, which I witnessed this morning – in which the induction coils and receiving magnets, as modified by these gentlemen, were made to actuate one of my recording instruments – have most satisfactorily resolved all doubts of the practicability as well as practicality of operating the telegraph from Newfoundland to Ireland.

Although we telegraphed signals at the rate of two hundred and ten, two hundred and forty-one, and according to the count at one time, even of two hundred and seventy per minute upon my telegraphic register, (which speed, you will perceive, is at a rate commercially advantageous,) these results were accomplished notwithstanding many disadvantages in our arrangements of a temporary and local character – disadvantages which will not occur in the use of our submarine cable.

Having passed the whole night with my active and agreeable collaborators, Dr. Whitehouse and Mr. Bright, without sleep, you will excuse the hurried and brief character of this note, which I could not refrain from sending you, since our experiments this morning settle the scientific and commercial points of our enterprise satisfactorily.

With respect and esteem, your obedient servant,

Samuel F. B. Morse

Field required further confirmation that the bed of the Atlantic Ocean was suitable for an underwater cable. For instance, he had to be sure that the ocean floor was of a suitable consistency and that no huge underwater mountains or valleys would interfere with the project. He called upon the U.S. Government, which commissioned the U.S. Navy to conduct another survey. It was carried out in July 1856 by the *Arctic*, which took measurements of the ocean bottom every forty miles or so. Lieutenant Berryman was again in charge and the survey corroborated the findings made several years earlier. Field was not yet totally convinced, so before proceeding with the project, he also requested the British Admiralty to conduct a similar survey. The Admiralty steamer *Cyclops* was dispatched to examine the route between Ireland and Newfoundland. This ship, under the command of Lieutenant Commander Joseph Dayman, conducted a more detailed survey than the *Arctic*, testing the ocean bottom every twenty or thirty miles. The survey results from both ships supported the earlier findings and Field was now satisfied installation of the cable could begin.

In its rush to complete the project in 1857, Field's company, Atlantic Telegraph hastily proceeded with the purchase of submarine cable. It placed an order with a value of £100,000 with the Gutta Percha Company (founded in 1845) to construct the main conductor. The conductor consisted of seven strands of No. 22 copper wire insulated with three layers of gutta-percha.

Once the insulated conductor was manufactured, it had to be made suitable for underwater use by enclosing it in an iron sheath. Half the conductor was taken by Glass, Elliot, & Company of East Greenwich and the other half by Newall & Company of Birkenhead. At both factories the conductor was covered in a metallic sheath consisting of eighteen strands of seven-strand iron wire. The thickness of the finished cable was five-eighths of an inch and its weight was 2000 pounds per mile. Unfortunately, the specifications for the cable were not clear and the factories wound the sheath on their part of the conductor in opposite directions, one from right to left and the other from left to right. Although this caused trouble when the cables were spliced together, the problem was overcome after some difficulty. Manufacture of the cables was completed in July of 1857.

THE CABLE SHIPS

To allow for slack and breakages, 2500 miles of cable were manufactured for the project. The cable weighed two and a half thousand tons, so any ship capable of carrying it had to be of considerable size. Except for the *Great Eastern*, still under construction, there was no ship that could carry the entire length of cable. It was therefore necessary to use two vessels. Both the British and American governments supplied ships for the project; the British government provided the battleship *Agamemnon* and the United States the *Niagara*.

The *Agamemnon* was 195 feet long, 54 feet wide and had a displacement of about 3500 tons. She was primarily a sail-powered ship, but also had an auxiliary 600 horsepower steam plant to power twin screws.[23] As a warship, she had previously seen action at Sebastopol during the Crimean War. The *Niagara* had a displacement of 5200 tons and was built as a steamship with auxiliary sail.[24] Her designer was George Steers, who also designed the sail ship *America* of the America's Cup race fame. Both ships were extensively modified for the cable laying project. The *Agamemnon* was accompanied by the paddle-steamer *Leopard* and the survey ship *Cyclops*. The *Niagara* was accompanied by the paddle-steamer *Susquehanna*.

It was decided that the eastern terminus would be Valentia, Ireland, where the cable would connect into the British

"Machines covering with gutta percha the Atlantic cable wire, at the Gutta Percha Company's works, Wharf-road," from the Illustrated London News, *14 March 1857*

and European telegraph systems. The western terminus would be Bull Arm (now Sunnyside), Trinity Bay, Newfoundland, where it would connect into New York Telegraph's lines to the rest of North America.

There was considerable debate between the electricians and engineers about how the cable should be laid. The engineers thought that the *Niagara* and *Agamemnon* should begin at mid-Atlantic where the two cables would be spliced. Each ship would then start laying the cable in opposite directions – the *Niagara* to Newfoundland and the *Agamemnon* to Ireland. The electricians, however, wanted to begin the cable from Valentia, with the *Niagara* starting the installation, and the *Agamemnon* splicing into *Niagara's* cable at mid-ocean. This would allow the cable to be continuously tested from Valentia as it was being laid. On this issue, the electricians got their way, and after taking several weeks to load the cable, the two ships headed to Ireland.

Cross section of the 1857 cable (Courtesy of the Heart's Content museum)

THE INSTALLATION BEGINS

At Valentia, the *Niagara's* cable was spliced into a larger shore cable connected to the cable station. On 7 August 1857, the *Niagara*, with the *Agamemnon* not far behind, headed west toward Newfoundland, laying the cable along the route.

After only five miles from Valentia, the cable became entangled in the *Niagara's* paying out equipment and broke away from the ship. The *Niagara* returned to Valentia, and after making a new splice into the shore cable, left the next day to continue the mission. By 9 August, it had successfully laid 95 miles of cable; however, on the next day, when the ship was about 385 miles from Valentia, the cable again broke, in about 2000 fathoms of water.[25] The *Niagara* had no choice but to return to port. As a considerable length of cable had been lost, it was too risky to restart the attempt with the remaining amount, so the mission was postponed until the following year.

Field began to realize that the Atlantic was not going to be conquered easily.

"View in Valentia Bay – The laying of the cable," *from the* Illustrated London News, *22 August 1857*

"The Atlantic Cable, ready for shipment, Morden Wharf, East Greenwich," from the Illustrated London News, *14 March, 1847.*

"The H.M.S. Agamemnon, 91 guns, shipping the English portion of the Atlantic submarine cable at East Greenwich," from the Illustrated London News, 1 August 1857

"Paying out the land end of the cable from the stern of the Niagara," from the Illustrated London News, 22 August 1857

*The cable station at Bull Arm (Sunnyside) 1857 – 1858
(from a drawing by Robert Dudley)*

Five

SUCCESS AND DISAPPOINTMENT – THE 1858 ATLANTIC CABLE

Field immediately began plans for another transatlantic cable attempt. After losing approximately £100,000 on the failed 1857 expedition, many of the company's investors were hesitant to continue with the project. However, the directors of Atlantic Telegraph were still optimistic the project would succeed and ordered another 900 miles of cable from Glass, Elliot, & Company. Field again approached the British and American governments, who continued to support the project by again making the *Agamemnon* and the *Niagara* available.

Field also obtained from the United States government the services of William Everett, who had impressed him as chief engineer on the *Niagara* during the 1857 attempt. After joining Field's team, Everett made design changes and implemented improvements to the equipment on both ships to provide better control over the strain on the cable as it was being laid. This "paying out" equipment was critical to the cable installation process because it controlled the speed at which the cable was laid. If the installation was too slow, the cable might break. On the other hand, if the installation was too fast, excessive cable would be used. A change was also made to the way the cable was stowed on the ships. Large cones were installed, around which the cable was wound rather than lying loose in the holds, thereby lessening the risk of it becoming entangled.

THE PROJECT BEGINS AGAIN

Since the electricians' earlier plan of laying the cable directly from Valentia to Newfoundland had failed, it was decided that the advice of the engineers would be followed this time. The *Agamemnon* and *Niagara* were to meet midpoint in the Atlantic, and begin laying the cable in both directions – the *Agamemnon* toward Valentia, Ireland in the east and the *Niagara* to Trinity Bay, Newfoundland in the west. On 10 June 1858, both ships departed Plymouth, England to meet at Latitude 52 2' and Longitude 33 18'. En route, a huge storm came up lasting more than a week. For a while it looked like the mission might have to be cancelled. Both ships were dangerously unstable because of the heavy weight of cable in their holds. Yet though they sustained considerable damage, they were able to continue the mission. Nicolas Wood was a correspondent with *The Times* on the *Agamemnon* and gave this dramatic first hand account of the storm:

> Our ship the *Agamemnon*, rolling many degrees – not everyone can imagine how she went it that night – was labouring so heavily she looked like breaking up.
>
> The massive beams under her upper deck coil (of cable) cracked and snapped with a noise resembling that of small artillery, almost drowning the hideous roar of the wind as it moaned and howled through the rigging, jerking and straining the little storm-sails as though it meant to tear them from the yards. Those in the impoverished cabins of the main deck had little sleep that night, for the upper deck planks above them were "working themselves free" as the sailors say; and beyond a doubt they were infinitely more free than easy, for they groaned under the pressure of the coil with a dreadful uproar, and availed themselves of the opportunity to let in a little light, with a good deal of water at every roll. The sea, too, kept striking with dull, heavy, violence against the vessel's bows, forcing its way through hawse holes and ill-closed ports with a heavy slush; and thence, hissing and winding aft, it aroused the inhabitants of the cabins aforesaid to the knowledge that their floors were under water, and that the flotsam and jetsam noises they heard beneath were only caused by their outfit for the voyage taking a cruise of its own in some five or six inches of dirty bilge. Such was Sunday night and such was a fair average of all the nights throughout theweek, varying only from bad to worse.

During the *Agamemnon's* harrowing experience, 45 crew members were injured, including some burned by flaming coal and boiling water. In addition, the cable on the main deck shifted and became entangled. This required several days to straighten out.

"The Agamemnon *in a storm," from the* Illustrated London News, *31 July 1858*

"The Atlantic telegraph paying-out machine," from the Illustrated London News, *3 July 1858*

By 25 June, things were put back in order and the two ships were ready to begin laying the cable. On 26 June, the cables from both ships were spliced,but before they moved off, the cable broke, and another splice was necessary. Once the new splice was completed, the ships headed in their respective directions. After the *Agamemnon* had laid out 37 miles of cable, and the *Niagara* 43 miles, the cable again broke. On 28 June, the ships met once again to effect a third splice. On 29 June, after placing more than 250 miles of cable, the cable again failed and the mission was halted. After two and some half weeks of storms and repeated failures, the ships headed back to port to assess their position.

ANOTHER TRY

Field was determined that the project should continue. After ensuring there was sufficient cable to complete the mission, the directors of Atlantic Telegraph ordered another try. This was after much debate and the resignation of the chairman and vice-chairman of the board of directors.[26] After two failed attempts, optimism on the success of the project was subsiding and there was little of the enthusiasm associated with the previous voyages. The ships replenished their food, coal and other supplies, and on 17 July quietly left Queenstown, Ireland to meet at mid-ocean.

On 28 July 1858 the *Agamemnon* and *Niagara*, each with 1100 nautical miles of cable on board, met at their rendezvous. On the next day they completed a splice and began heading to their respective destinations.

From the *Niagara's* navigation reports on 29 July, the ship was 16 miles off course. If this deviation continued, there would not be sufficient cable to complete the project. Upon investigation, it was discovered that the massive bulk of steel in the cable was affecting the ship's compass. It was therefore decided that the *Niagara* would have to be guided by its escort, the *Gorgon*, which had a properly working compass. Apart from the compass problem, the voyages of the *Niagara* and *Agamemnon* were uneventful. On 30 July, 265 miles of cable were paid out; by 31 July, 540 miles; by 1 August, 884 miles; by 2 August, 1256 miles; by 4 August, 1854 miles; and by the time the *Agamemnon* arrived at Valentia on 5 August, 2022 miles. The *Niagara* arrived at Bull Arm, Trinity Bay on the evening of 4 August, and was met by the *Porcupine*, which had earlier arrived to make arrangements for the shore end of the cable. From a lookout on a nearby hill, watchers saw the *Niagara* when it was more than 20 miles from shore, and signalled the *Porcupine* with a fire. The *Porcupine* met the *Niagara* in Trinity Bay and guided it to the cable landing site. The following day, the cable was landed at Bull Arm and was connected to the cable station two miles inland. Earlier that day, the *Niagara* received a message from the *Agamemnon* that it had arrived at Valentia.

Cyrus Field addressed the assembled crowd and told them that he had established contact with Valentia. He wasted no time in telegraphing news of the transatlantic cable landing to President Buchanan of the United States and to the Mayor of New York. He also sent a message to the Associated Press:

United States Frigate Niagara
Trinity Bay, Newfoundland, August 5, 1858.

To the Associated Press, New York:

The Atlantic Telegraph fleet sailed from Queenstown, Ireland, Saturday, July seventeenth, and met in mid-ocean Wednesday, July twenty-eighth. Made the splice at one PM Thursday, the twenty-ninth, and separated – the *Agamemnon* and *Valorous*, bound to Valentia, Ireland; the *Niagara* and *Gorgon*, for this place, where they arrived yesterday, and this morning the end of the cable will be landed.

It is one thousand six hundred and ninety-six nautical, or one thousand nine hundred and fifty statute, miles from the Telegraph House at the head of Valentia harbour to the Telegraph House at the Bay of Bulls, Trinity Bay, and for more than two thirds of this distance the water is over two miles in depth. The cable has been paid out from the *Agamemnon* at about the same speed as from the *Niagara*. The electric signals sent and received through the whole cable are perfect.

The machinery for paying out the cable worked in the most satisfactory manner, and was not stopped for a single moment from the time the splice was made until we arrived here.

Captain Hudson, Messrs. Everett and Woodhouse, the engineers, the electricians, the officers of the ship, and in fact, every man on board the telegraph fleet, has exerted himself to the utmost to make the expedition successful, and by the blessing of Divine Providence, it has succeeded.

After the end of the cable is landed and connected with the land line of telegraph, and the *Niagara* has discharged some cargo belonging to the Telegraph Company, she will go to St. John's for coal, and then proceed at once to New York.

Cyrus W. Field

News of the transatlantic cable spread very quickly throughout Europe. Queen Victoria heard the news while she was dining with the Emperor of France in Cherbourg Harbour. She later knighted Charles Bright, the chief engineer of Atlantic Telegraph, for his engineering contribution to the project.[27]

After the transatlantic cable was landed, some of the ships sailed to St. John's, where festivities were held to celebrate their success. The ships' crews attended the St. John's regatta at Quidi Vidi Lake, where they were guests of honour. On 11 August, after loading on additional coal, the *Niagara*, along with Field and his entourage, headed to New York, where there were many celebrations honouring his feat. Newspapers all over the world praised Field, and a parade was held in New York City in his honour.

Field returned to New York satisfied that his task was completed. He wrote a letter to the Directors of Atlantic Telegraph in London, resigning as General Manager of the company, and had thoughts of continuing the retirement he had started several years before.[28]

There were however further surprises ahead.

OFFICIAL CONGRATULATIONS

The first few days of the transatlantic cable, as expected, were used for testing and congratulatory messages. On 16 August, Queen Victoria of England sent the following message to President Buchanan of the United States:

> The Queen desires to congratulate the President upon the successful completion of this great international work, in which the Queen has taken the deepest interest.
> The Queen is convinced that the President will join her in fervently hoping that the electric cable which now connects Great Britain with the United States, will prove an additional link between the two nations, whose friendship is founded upon their common interest and reciprocal esteem.

The Queen has much pleasure in thus communicating with the President, and in renewing to him her best wishes for the prosperity of the United States.[29]

On 16 August 1858 President Buchanan responded to Queen Victoria:

The President cordially reciprocates the congratulations of Her Majesty the Queen on the success of the great international enterprise accomplished by the science, skill, and indomitable energy of the two countries.

It is a triumph more glorious, because far more useful to mankind, than was ever won by conqueror on the field of battle.

May the Atlantic Telegraph, under the blessing of Heaven, prove to be a bond of perpetual peace and friendship between the kindred nations, and an instrument destined by Divine Providence to diffuse religion, civilisation, liberty, and law throughout the world.

In this view will not all nations of Christendom spontaneously unite in the declaration that it shall be for ever neutral, and that its communications shall be sacred in passing to their places of destination, even in the midst of hostilities.[30]

The Board of Directors of Atlantic Telegraph were also among the first to use the new cable. On 17 August, they sent the following message to Field:

The Directors have just met, they heartily congratulate you on success. *Agamemnon* arrived at Valentia at 6 am Thursday August 5. We are just on the point of chartering ship to lay shore end. No time will be lost in sending them out. All your letters have been posted to New York. Please write to me fully about tariff and other working arrangements.[31]

It is obvious from this message that the eastern end of the cable was still not connected to the cable station at Valentia, but this was completed several days later.

One of the first public messages was transmitted over the cable on 17 August 1858 when the Cunard Shipping Line sent a message to London advising that no lives were lost in the collision of the *Arabia* and *Europa*, after both ships had arrived safely in St.

John's.[32] Several weeks later, in what was to be one of the last messages sent over the cable, the British government ordered its garrison in Nova Scotia to remain in place, after it had previously ordered it home by steamer mail to provide support in an uprising in India. The telegraph message reached the regiment shortly before they were ready to depart, saving the British government tens of thousands of pounds.

THE CABLE FAILS

Unfortunately, the quality of the cable's signal deteriorated quickly and only 400 messages were handled during the first three weeks. By 3 September 1858 the cable could no longer be used to transmit commercial messages, and by 20 October it had gone completely dead. After the euphoria of the earlier success, the public had become sceptical of the viability of such a difficult engineering feat. Field was for a short while an American hero, but after the cable's failure, people called him a swindler and some even made accusations that the transatlantic cable was a gigantic hoax.[33]

Atlantic Telegraph's engineers and electricians debated why the cable had failed. Some suggested that the cable had been manufactured too hastily and was not properly tested. Others thought that the cable was damaged by the ships' paying out and braking equipment. Some maintained that the repeated coiling and uncoiling of the cable had damaged the conductor. However, the general consensus was that the cable was irreparably damaged by applying too much voltage and burning out the insulation.

The Company's chief electrician was Dr. Edward Whitehouse, a medical doctor by profession, who was keenly interested in electricity and responsible for designing the apparatus for receiving and transmitting the signals over the cable. Whitehouse initially used ordinary telegraph instruments, which did not work satisfactorily because of the distance involved. He subsequently considerably increased the voltage on the line, and in the opinion of many, applied too much power to the cable and burnt out its insulation. Although he strongly denied responsibility, Whitehouse was fired. Whether he was responsible will never be known, but the cable could not be restored, and several attempts to recover it for repair or future use were not successful.

Although the viability of a transatlantic cable was proven, the project had failed, and the value of Atlantic Telegraph's shares plummeted. Field had to reassess his position. His company had lost £500,000 and plans for another cable looked bleak. For a couple of years he considered another attempt, but the American Civil War broke out in 1861, interfering with any immediate plans. For the time being, thoughts of a transatlantic cable were put on hold and Field had to find satisfaction in selling surplus cable from the failed expedition to Tiffany and Company, a prominent jewellery house in New York, which sold six-inch remnants of the cable, mounted in silver, as expensive souvenirs.

"VIA CAPE RACE"

Although the transatlantic cable had failed, news from Europe was still important to large American newspapers. Competition for European news was so great that major newspapers employed extraordinary means to ensure their papers were first with the news. In Nova Scotia, for instance, the Associated Press had used a pony express to carry news from ships arriving at Halifax to the nearest telegraph line in the United States. The pony express was later disbanded when Halifax was connected to the North American telegraph system.

"Cape Race, Newfoundland, the termination of the American system of telegraphs," from the Illustrated London News, *24 August 1861*

To further increase the speed of news from Europe, the Associated Press stationed a boat at Cape Race in 1859 to intercept transatlantic ocean steamers passing nearby on their way to Halifax and New York. John Murphy was the captain of the pickup boat. Watertight canisters containing news and messages from Europe were thrown overboard from the steamers, retrieved by the pickup boat and telegraphed to North American newspapers from Cape Race. This was no trivial exercise, as the sea off the cape is famous for its unpredictability, with no nearby harbour to find refuge. The first ocean steamer to drop off news at Cape Race was the *Vigo* of the Inman line. Collecting news this way was immortalized in North American newspapers with the byline "Via Cape Race," and continued until the first commercially successful transatlantic cable was completed by Cyrus Field in 1866.

Six

OTHER PROPOSALS

Field was not the only person with ambitions to connect America and Europe by telegraph. Earlier, the views of Bishop Mullock and Samuel Morse on a transatlantic cable were mentioned. In the 1840s, others took the idea a step further and announced plans for such a venture. In 1846, Jacob Brett registered a company in England called the General Oceanic and Subterranean Electric Printing Telegraphic Company. The objective of this company was "To establish a telegraphic communication from the British Islands across the Atlantic Ocean to Nova Scotia and Canada and establishing electric communication by land and sea with the Colonies."[34]

In 1847, Samuel Armstrong, an American manufacturer, laid a cable insulated with gutta-percha across the Hudson river in New York. Impressed by this early success, he proposed in the *New York Journal of Commerce* that a cable could be placed across the Atlantic Ocean. He estimated that the cost of such a project would be about $3.5 million.[35]

Other proposals for transatlantic communications included a cable across the South Atlantic by way of Spain, Madeira, the Canary Islands, Cape de Verde Islands, Don Pedro, Fernando de Noronha Isles to the West Indies and the United States. Another plan called for a cable from Portugal, to the Azores, then to Bermuda and the United States. None of these plans found favour, especially since they had no support from Britain, which was then the political and economic centre of Europe.

"The telegraph expedition company camping in Labrador," from the Illustrated London News, *7 July 1860*

At the time, such undertakings were innovative ideas, and for whatever reasons, they proceeded no further. It would be almost a decade later before the idea would arise again.

COLONEL SHAFFNER

After Field's failure with the 1857 and 1858 cables, most investors were reluctant to consider another transatlantic undertaking. However, there were other bold promoters undeterred by Field's misadventures who thought they could succeed where he had failed. One such individual was Colonel Taliaferro P. Shaffner, an American businessman who had built several telegraph systems in the United States. Shaffner believed that the 1858 underwater route was far too lengthy to render a transatlantic telegraph system practical and would never provide reliable communications. The cost of messages would always be prohibitive, making the cable uneconomic. He therefore proposed a route from Labrador to Greenland, Iceland, the Faeroe Islands, Norway, Denmark and the rest of Europe. The longest underwater section of Shaffner's proposed route was about 600 miles, so the cable could be operated more efficiently and economically. Shaffner later changed the route to bypass Denmark, and go directly from Norway to Scotland. He obtained the approval of the governments involved and stirred up considerable interest in Britain and Europe. The British government was anxious to see a telegraph system to the United States and its North American colonies and in 1860 conducted submarine surveys of the proposed route. The survey was conducted by the *Bulldog* and *Fox*.[36]

Despite the preliminary work, which included his personal tracing of the route in his own ship, Shaffner's plan proceeded nowhere, probably because of his inability to raise funds to finance such a project. There were also the technical concerns of maintaining a cable so far north, not to mention the routing of the proposed cable, which crossed areas of ocean where iceberg activity was very high.

COLLINS OVERLAND TELEGRAPH

A few years later, another proponent of a telegraph cable between North America and Europe surfaced. This time however, the route was via the Bering Strait rather than the North Atlantic. The promoter was an American by the name of Perry McDonough Collins (1813-1900), an entrepreneur who at one time served as a U.S. commercial agent to Russia. Collins found himself in California during the gold rush where he promoted several successful business ventures. During his visits to Russia in the late 1850s, he became convinced that a tremendous trade opportunity existed between the United States and Russia, especially if communications between the two countries could be improved. Since Field's 1858 transatlantic cable had failed, an Atlantic telegraph did not seem feasible, so Collins came up with the idea of connecting North America telegraphically with Europe via Asia. His plan was to extend Western Union's lines up the west coast to British Columbia, through the Yukon and Alaska (which was then Russian America), across the Bering Strait, through Siberia to Russia and then eventually on to Europe. Collins approached Western Union, who gave him a $5000 grant to seek approval from the Russian and English governments. He obtained the agreement of Russia in May 1863 for a right-of-way from the Bering Strait to Amur, the location of its easternmost telegraph line. He also obtained approval of the British government for a right-of-way through British Columbia and the Yukon.

Collins submitted his report to Western Union in September 1863. In March 1864, the company offered to purchase his rights in the project for $100,000 cash and 10,000 shares of its stock, which at the time had a par value of $1,000,000. Collins accepted, but continued his involvement in the project as well as being made a member of Western Union's board of directors.

The route through Russia was surveyed, and with the agreement of the governments involved and the financial backing of Western Union, the project looked promising. Construction on the Collins Overland Telegraph started in late spring of 1865, and by the summer of 1866 was well under way. A pole line had been extended from the state of Washington, through British Columbia to the Yukon. In Russia, construction was even further advanced with the construction of several thousand miles of pole lines.

On 27 July 1866, however, Field successfully landed a workable transatlantic telegraph cable. Nonetheless, Western Union was not convinced that a transatlantic cable would be viable, and continued work on the route through Asia. In February 1867, however, they issued a stop work order, finally convinced that the transatlantic cable would be a permanent success. Western Union lost more than three million dollars on the project. Its only beneficiaries were Collins, who was paid cash and shares by Western Union, and Russian officials, who were offered 1000 shares of Western Union to ensure their cooperation. The United States government also benefited by recognizing Alaska's strategic importance and eventually agreeing to purchase Alaska from Russia for the bargain price of about two cents an acre.[37]

Seven

SO CLOSE BUT YET SO FAR – THE 1865 ATLANTIC CABLE

Despite the failure of the 1858 cable, its initial success convinced Field that another attempt should be made. He was further encouraged by the British government which was interested in linking its far-flung empire together by electric telegraph. By this time, there were several long underwater cables in service, including cables across the Mediterranean and the Persian Gulf. Despite losing a considerable amount of money in its failed Red Sea underwater cable, the British government continued to provide support to Field. It increased its subsidy to Atlantic Telegraph to £20,000 a year and guaranteed a fixed return on new capital for a period of twenty-one years.[38] Convinced that a transatlantic cable would eventually succeed, the government ordered new soundings of the ocean bottom and insisted on extensive testing of the new cable before installation.

THE TECHNICAL COMMITTEE

In 1859 the British Board of Trade set up a committee of Britain's top scientists and engineers to look into the technical details of a transatlantic telegraph cable. The committee sat for more than two years, during which time it thoroughly investigated all aspects of submarine telegraphy, including matters relating to the cable, such as size, weight, flexibility, insulation and electrical qualities. One of the committee's findings was that the high voltage used on the 1858 cable was not necessary and that a lower voltage would be sufficient, reducing the risk in breaking down the cable's insulation. After extensive work, the committee issued a comprehensive report. Their conclusions were summarized in the following certificate, which was signed by all involved.[39]

London, 13th July, 1863

We, the undersigned, members of the Committee, who were appointed by the Board of Trade, in 1859, to investigate the question of submarine telegraphy, and whose investigation continued from that time to April, 1861, do hereby state, as the result of our deliberations, that a well-insulated cable, properly protected, of suitable specific gravity, made with care, and tested under water throughout its progress with the best known apparatus, and paid into the ocean with the most improved machinery, possesses every prospect of not only being successfully laid in the first instance, but may reasonably be relied upon to continue for many years in an efficient state for the transmission of signals.

Douglas Galton,	C. Wheatstone,
Cromwell F. Varley,	Latimer Clark,
William Fairbairn,	Edwin Clark,
George P. Bidder,	George Saward

RAISING THE CAPITAL

Field estimated it would cost approximately £600,000 to purchase and install a new transatlantic cable. Because the United States was embroiled in a civil war, it was difficult to raise capital there, so he could only obtain £285,000 in that country. The remaining amount was raised in Britain, where he found the required capital with the Telegraph Construction and Maintenance Company.

The Telegraph Construction and Maintenance Company (Telcon) was registered on 7 April 1864 with an authorized capital of £1,000,000. The company was set up by Glass Elliot & Company and the Gutta Percha Company, the manufacturers of the 1858 cable. These companies had manufactured other long underwater cables, including the cable across the Mediterranean connecting Malta and Alexandria, as well as the 1500 mile long cable across the Persian Gulf, enabling direct communications between Great Britain and India. The major shareholders of Telcon were Thomas Brassey, a capitalist, John Pender, a member of Parliament, John Chatterton, who invented an insulating material, Willoughby Smith, an electrician, and R. A. Glass, a director of Glass Elliot & Company. The company agreed to purchase £315,000 worth of shares in Atlantic Telegraph and provide £100,000 in bonds provided it was awarded the contract to lay the transatlantic cable. This offer effectively gave Telcon control over the undertaking. Atlantic Telegraph accepted Telcon's proposal and quickly began work on organizing the project.[40]

THE NEW CABLE

The new cable was manufactured to the highest standards. As recommended by the submarine technical committee, every part of the cable was put through the most stringent of tests before the project go-ahead was given. Seven strands of copper of very high purity formed the main conductor, which was three times larger than the 1858 conductor. The cable also had improved insulation and far better electrical and mechanical properties. It measured 1.1 inches in diameter and weighed almost two tons per mile. The cable had twice the tensile strength and was easier to lay than the earlier one. It was 2,300 nautical miles in length and was manufactured at the rate of fourteen miles per day, requiring about eight months to complete.

Based on the additional hydrographic surveys, it was decided that a new location in Newfoundland would be more desirable than Sunnyside, the site of the 1858 cable. The new port selected was Heart's Content, Trinity Bay, which in addition to having a shorter route, had an ideal harbour for landing the new cable.

THE GREAT EASTERN

The new cable was more than twice as heavy as the one used in 1858. Furthermore, it was planned to lay the cable in one continuous run from Valentia to Newfoundland. What ship could manage such an immense cargo? The *Agamemnon* and the *Niagara* were far too small; in fact, both ships in total could barely carry half the cable for the project. There was only one ship afloat which could carry such a huge load – the *Great Eastern*.

The *Great Eastern* was designed by Isambard Kingdom Brunel, one of the foremost engineers of the nineteenth century. Included with the *Great Eastern* in his list of credits were the first iron ship, a number of European bridges, the Great Western Railway and the first screw-propeller ship to cross the Atlantic. Brunel's engineering projects were usually large scale, and the *Great Eastern* was certainly no exception. Field was given a personal tour of this great ship by its design engineer while it was under construction and probably wished that the ship had been available for his earlier transatlantic projects.

The ship was launched in 1858 and cost £640,000 to build. It was initially launched as *Leviathan*, but its name was subsequently changed to *Great Eastern*. It was 693 feet in length, 120 feet wide with a displacement of 22,500 tons, five times greater than any existing ship.[41] The *Great Eastern* was unusual not only for its huge size, but because it used three different propulsion systems. The ship was designed to work under sail power, a screw-propeller system and two paddle-wheels. It had six masts accommodating 6,500 square feet of sail as well as five smokestacks servicing two steam engines: the first, a 4,890-horsepower engine providing power to the screw-propeller, and the second, a 3,410-horsepower engine serving the paddle-wheels. The *Great Eastern*, without sail power, and with paddle-wheels and screw-propeller working in unison, could move at fifteen knots. The ship had a capacity of up to 4000 passengers and for a while was used as a transatlantic passenger liner. It was, however, too large for the route and its operation for this purpose was not financially viable. It became a commercial failure and its owners were quite anxious to dispose of their investment. The *Great Eastern* was therefore put up for sale.

ENTER DANIEL GOOCH

Atlantic Telegraph did not have the resources to purchase and refit the *Great Eastern*; however, Daniel Gooch, a Director of the company, was interested in making an offer. Eventually, Gooch and his partners purchased the *Great Eastern* for only £25,000, an amount which obviously suggests that few thought that the ship had a commercial future.

Confident that a transatlantic telegraph would succeed, Gooch offered the *Great Eastern* to lay the cable, provided he receive £50,000 in stock of Telcon if the venture were successful. His offer was accepted, and with arrangements for the *Great Eastern* now secure, the project could continue.

Since the *Great Eastern* was originally a passenger ship, extensive modifications were required to make it suitable for the task, including the installation of cable holding tanks and cable laying equipment. The ship's number four smokestack, as well as several saloons and cabins, were removed to allow additional room for cable storage. Three holding tanks were installed: one afore ship measuring 52 feet in diameter and 21 feet deep, capable of holding 693 miles of cable; another mid ship, 59 feet in diameter and 21 feet deep, holding 899 miles; and the third after ship, 58 feet in diameter and 21 feet deep, holding 898 miles. The modifications took almost half a year to complete. Once the ship was ready, the cable was loaded at Sheerness, about 35 miles southeast of London. The cable was transported to the *Great Eastern* by the bulk carriers *Amethyst* and *Iris* at the rate of about 22 miles per day. It was coiled into the holding tanks filled with water so that it would not dry out. Approximately 2,300 miles of cable were loaded on the ship, which was about 700 miles more than the length required between Ireland and Newfoundland. The cable weighed about 5,000 tons and the water in the holding tanks weighed another 2,000 tons. The paying out equipment on the *Great Eastern* was designed by Samuel Canning of Telcon and was manufactured by Penn & Co. of Greenwich.

The Prince of Wales visited the *Great Eastern* on 24 May 1865. He inspected the ship and witnessed a test of the 1400 mile portion of cable that was already loaded. On 15 July 1865, the *Great Eastern* departed Sheerness for Ireland, captained by James Anderson, who had years of experience sailing the Atlantic with the Cunard Line. Assisting him was Captain Moriarity, seconded from the British Admiralty, who had served on the *Agamemnon* during the previous attempt. Samuel Canning was in charge of the cable laying operation, and was assisted by C. V. de Sauty who was in charge of the electricians. Atlantic Telegraph was represented by Mr. Varley, its chief electrician, and Professor Thompson, both of whom were responsible for ensuring that conditions of the contract were met. Cyrus Field, Daniel Gooch and Charles Bright were also on board for the crossing.

There was a clamour from members of the press to accompany the voyage. Telcon did not want reporters on board and banished them from the ship. After recognizing the significance of the project, the company relented, and agreed that the venture should be recorded for historical purposes. The company allowed William Howard Russell, a correspondent with *The Times*, to accompany the voyage. Russell had earlier gained fame as an intrepid war correspondent, reporting on the Crimean war, the Franco-Prussian war and the US civil war. He was accompanied by artists Robert Dudley of the *Illustrated London News*, and Henry O'Neill to record the venture.[42]

The *Great Eastern's* mission was by no means an ordinary one, and a large amount of supplies was required. Besides its cargo of cable, the ship carried almost 500 men, 8000 tons of coal, twelve oxen, one cow, and twenty pigs.[43]

The *Great Eastern* steamed for Valentia in Foilhummerum Bay, Ireland. The shore landing cable had been laid earlier by the *Caroline*, and despite high winds, the transatlantic section was spliced into it, making a connection to the cable station. The station was previously connected by telegraph lines to Killarney and the European telegraph system. On 23 July 1865, the *Great Eastern*, escorted by the *Sphinx* and the *Terrible*, turned west and headed toward Heart's Content, Newfoundland.

Under Canning's direction, the cable-laying proceeded smoothly. The electricians tested the cable as it was being laid and maintained communications with Valentia using Morse code. The *Great Eastern* was sent the latest news, including a message to Gooch advising him that he had been elected to the British parliament. Obviously a better businessman than parliamentarian, Gooch sat in four parliaments and apparently didn't speak a word during his tenure.

THE PROJECT BEGINS

The status of the expedition was reported to project officials in London. Things were going well until 84 miles from Valentia, when de Sauty's electricians detected a fault in the cable. Using a Wheatstone bridge, the engineers measured the resistance of the cable to determine the distance the fault was from the ship. By using different methods of calculations, they estimated that it was located more than ten miles from the *Great Eastern*. Subsequently, the cable was cut and transferred to the bow of the ship, where it was connected to a winch and rolled back in as the ship retraced its course. As it was being wound in, the cable was closely examined, and after ten and a half miles were recovered, the fault was found. It was caused by a two-inch piece of metal which had pierced the cable and shorted out the conductor. After making repairs, the cable was transferred to the paying out equipment, and the ship again turned west toward Newfoundland.

The expedition progressed smoothly, with the *Sphinx* performing soundings ahead of the *Great Eastern* while the *Terrible* watched out for shipping in the area. Cable testing with Valentia proceeded satisfactorily and the results were much better than predicted. However, on 29 July, after more than 700 miles of cable had been laid, another fault occurred. Again the cable was rolled back in, and to everyone's horror, the fault was again caused by a piece of iron piercing the cable. The engineers thought that two faults caused by the same problem were unlikely to happen by chance. As the source of the faults looked suspicious, sabotage was suspected. Canning therefore implemented strict security, and only those directly involved with laying the cable were allowed near it. By 30 July, the cable had been repaired, and in a thick fog, the *Great Eastern* again headed toward Trinity Bay.

By 31 July, all the cable in the *Great Eastern's* after-hold was laid, and the crew switched to the cable in the fore-hold. All on board were looking forward to a smooth and successful expedition. However, on 2 August, almost 1200 miles from Valentia, another defect was discovered. This time the fault was noticed before the cable was paid out. To everyone's horror, it was again caused by a piece of iron. However, the mystery of the earlier incidents now became clear. The engineers discovered that the faults were caused by pieces of broken iron sheath which had fallen deep into the hold, and contrary to Canning's earlier concern, were not caused by sabotage.

Unfortunately, the ship could not be slowed in time to stop the faulty cable from reeling into the sea. The ship finally stopped and the cable was transferred from the stern to the bow to be wound back in. This time however, disaster struck. The tension on the cable suddenly increased, caused either by the rolling of the ship or a mistake with the cable recovery equipment. The cable snapped, broke away from the ship and sank more than two miles to the bottom of the ocean.

In anticipation of such problems, the *Great Eastern* was equipped with grappling equipment to retrieve lost cable. Under Canning's direction, the ship plied back and forth over the cable trying to hook it with its grappling gear. It was finally hooked, and the reeling in began. However, the combination of its weight and great depth were too much for the grappling equipment to handle. The wire rope snapped, and the rope, grapnel hook and cable sank back into the Atlantic. Two other grappling attempts were attempted but both were unsuccessful. There was no other option but to halt operations. The Atlantic remained unconquered, and once again another transatlantic cable attempt had failed.

Cyrus Field had suffered another severe setback. Reporting on the failure, W. H. Russell, the *Times* correspondent on the *Great Eastern*, noted Field's disappointment in the following dispatch:

> Mr. Field came from the companionway into the saloon, and observed with admirable composure, though his lips quivered, and his cheek was white: 'The cable has parted and has gone from the reel overboard.'

The *Great Eastern* marked the location of the cable with buoys, and on 12 August, a disappointed Field and entourage steamed back to England.

"Landing the Atlantic cable in Heart's Content Bay, Newfoundland,"
from the Illustrated London News, *8 September 1866*

Eight

TRIUMPH AT LAST
–THE 1866 ATLANTIC CABLE

Although the failed 1865 venture had cost Field and his associates £600,000, they were determined to install a telegraph cable across the Atlantic. However, additional financing had to be obtained. Accordingly, Atlantic Telegraph began raising capital for the new project by offering preferred shares, all of which were quickly subscribed. However, the share issue was challenged on legal irregularities, because the company did not have the right to issue preferred stock. All the capital that was raised had to be returned and Atlantic Telegraph found itself back where it had started. Once again it was Daniel Gooch, the owner of the *Great Eastern*, who stepped forward to help the project.

Gooch proposed the formation of a new company which could legally raise capital for the enterprise, and then amalgamate itself with Atlantic Telegraph. The new company was set up, and shares were quickly subscribed, with Telcon the major shareholder. Telcon agreed to manufacture and place the new cable for £500,000 plus an additional £100,000 if the transatlantic cable were successful.[44]

The new company amalgamated with Atlantic Telegraph, and the resulting enterprise was named Anglo-American Telegraph (Anglo-American). With financing now in place, preparations immediately began on another attempt to span the Atlantic.

Anglo-American was confident of success and planned not only to install a new cable, but also raise the sunken cable from the previous attempt, splice into it and complete a second cable. The *Great Eastern* was thoroughly overhauled and a two-foot thickness of barnacles was cleaned off her hull, allowing the ship higher speed and greater manoeuverability. The ship was also fitted with improved paying out and recovery equipment. In the 1865 attempt, the recovery equipment was located at the bow, and the cable had to be cut and manually transferred from the stern. For the 1866 attempt, the *Great Eastern* was modified to allow it to conduct cable recovery operations completely from the stern, eliminating the difficult and risky method previously used. The ship was also loaded with more than 20 miles of strong grappling rope, capable of sustaining loads of up to 30 tons.

For this attempt, it was decided that a lighter and stronger cable than that used in 1865 would be required. It was manufactured by Telcon at the rate of 20 miles per day, and loaded on the *Great Eastern* at Sheerness. On 30 June 1866, the *Great Eastern* steamed toward Valentia to start the project. Without the fanfare and public excitement of the 1865 attempt, the shore end of the cable was landed by the *William Cory* and brought to the cable station at Valentia. On 13 July, after splicing into the shore cable, the *Great Eastern* headed west toward Heart's Content, Newfoundland. It was supported by the *Medway*, carrying an extra supply of cable, and the *Albany*, both of which were rigged out with grappling equipment. The *Terrible* and the *Raccoon* also accompanied the *Great Eastern*.

On this attempt the process of paying out the cable proceeded without incident. As in the case of the previous attempt, the ship kept in communication with Valentia as the cable was being laid. The ships reached the Grand Banks on 25 July, where they were slowed because of heavy fog. On 27 July, in a heavy mist, the *Great Eastern* arrived outside Heart's Content in Trinity Bay, where it was met by the *Niger*.

Field's dream of a transatlantic cable was soon to be realized, and from the *Great Eastern*, he wired the following message to Anglo-American's office in London.[45]

> *Great Eastern,* Heart's Content,
> 27th July, 1866.
> > We arrived here at 9 o'clock this morning. All Well.
> Thank God, the cable is laid, and is in perfect working order.
>
> > > > Cyrus Field

At Heart's Content, the anticipation of the *Great Eastern's* arrival was tremendous. Thousands of people from the surrounding area headed toward the town. Accommodation in the area was almost impossible to find, and people slept in barns, tents, or any other place that would provide shelter. Of all the visitors to the town, there was probably no one who received as much public attention as a young blind girl, who along with her brother walked from a nearby town to visit the *Great Eastern.*

"Laying the shore end of the cable from the Great Eastern, *in Heart's Content Bay, Newfoundland," from the* Illustrated London News, *8 September 1866*

While the *Great Eastern* was at anchor in Trinity Bay, the fog moved off, allowing the *Medway* to land the shore end of the cable. The sailors of the *Medway* carried the cable the last few feet to shore, up to their waists in water. In the meantime, the overseas and shore end cables were spliced together on the deck of the *Great Eastern.* Field was part of the landing party and waded ashore. The waiting crowd recognized him and cheered him for his historic accomplishment. Everywhere there were celebrations, and when the *Great Eastern* finally entered port, there was a salute of cannon and musket fire.

On arriving at Heart's Content, Field learned that the cable between Newfoundland and Nova Scotia was out of service, and messages would have to be relayed across the Cabot Strait by steamer. He immediately telegraphed St. John's and made arrangements for the steamer *Bloodhound* to proceed to the Gulf of St. Lawrence to make the repairs. Since this would take some time, he arranged for the *Dauntless* to carry messages across the Gulf until the cable was repaired. The *Bloodhound* loaded a supply of cable from the *Great Eastern* and proceeded to the Gulf.

At Heart's Content, Mr. Lundy of Atlantic Telegraph had earlier arrived from England and was in charge of making arrangements for landing the cable. He rented a house being built for James Legge and hired Legge, James Moore, Jonathan Hopkins, Thomas Jeans, along with J. H. Moore as foreman, to make the necessary renovations.[46] He also hired Alexander Smith, a stone cutter, to build a solid pedestal on which the mirror galvanometer would rest.

The mirror galvanometer was a sensitive receiving apparatus used in the early days of transoceanic telegraphy to pick up the weak transmission signals. It was invented by William Thompson, who was involved in the early development of telegraphy, and closely associated with Atlantic Telegraph. The device consisted of a small mirror suspended by a thread inside a coil of wire between two magnets. When an electric current from the cable was passed through the coil, the mirror would deflect, the amount of deflection depending on the magnitude of the current. A small lamp producing a beam of light reflected off the mirror onto a graduated scale two or three feet away. Small deflections in the mirror caused large changes on the graduated scale, making the movement of the mirror easier to read. When a positive signal was received, the galvanometer's light deflected to the right for a "dot." When a negative signal was received, the light deflected to the left for a "dash." The dots and dashes were Morse code for letters of the alphabet. Normally, one operator would monitor the message as it was received, and read this to a clerk who would write it down and pass it on to another for transmission to the next location. Different variations of the galvanometer were also used to measure electrical characteristics of the cable, making the device invaluable to engineers in performing maintenance and locating the location of faults and breaks.

In short order, the electricians connected the transatlantic cable with New York Telegraph's cable system to Harbour Grace, and trained operators from the *Great Eastern* prepared to handle messages to and from Europe. On 27 July 1866, Telcon handed over the new cable to Anglo-American in the presence of Gooch, Field, and Hamilton, Directors of Anglo-American, and Mr. J. C. Deane, secretary.[47] The station's first superintendent was the aforementioned Mr. Lundy. Other Heart's Content staff who arrived on the *Great Eastern* were:[48]

> Isaac Angel, clerk
> James Bartlett, clerk
> George Charlton, outdoor overseer
> Mr. Collett, traffic superintendent
> William Dickenson, electrician
> Frank Perry, clerk
> ·John Sullivan, accountant
> William Woodcock, clerk
> Charles Trippe,
> George Unicome, clerk
> Ezra Weedon, assistant superintendent

OFFICIAL MESSAGES BEGIN

The following message from Queen Victoria, from her summer residence at Osborne on the Isle of Wight, was received on the *Great Eastern* on 27 July and relayed from the Heart's Content station to President Johnson of the United States:

> From H.M. The Queen, Osborne.
> To the President of the United States, Washington.
>
> The Queen congratulates the President on the successful completion of the undertaking which she hopes may serve as an additional bond of union between the United States and England.

Because of problems with the Cabot Strait portion of the cable, it took some time for President Johnson to receive and respond to Queen Victoria's message. His response to the Queen was not received at Heart's Content and relayed to England until 31 July. President Johnson's reply message was as follows:

From the President of the United States,
Executive Mansion, Washington.

To the H. M. the Queen of the United Kingdom of Britain and Ireland.

 The President of the United States acknowledges with profound gratification the receipt of Her Majesty's message and cordially reciprocates the hope that the cable which now unites the Eastern and Western hemispheres may serve to strengthen and perpetuate peace and amity between the governments of England and the Republic of the United States of America.

 On Wednesday, 8 August, Newfoundland Governor Musgrave arrived from St. John's on the *Lilly* and was given a tour of the *Great Eastern*.[49] Once the cable was in operating order, company officials, ships' crew members and the general public commemorated the occasion in a service at the Heart's Content Church of England church. Reverend George Chandler presided over the service, while his daughter played the organ. This organ is presently on display at the Heart's Content cable station museum.

 On the matter of the blind girl (later identified as Mary Piercey) mentioned earlier, who had come to visit the *Great Eastern*, John C. Deane, Field's secretary, wrote in the *St. John's Public Ledger*:

 Saturday, July 28, 1866, ships alongside supplying us with coal and the accommodation ladder supplying us with visitors from all parts of Newfoundland, who braved the difficulties of the very worst roads and about as bad vehicles to get to Heart's Content to see the ship. But one came to whom it was not permitted to have that gratification, a blind girl, led by her young brother, walked about the deck and gathered from his intelligent description and by the exercise of her sense of touch, some notion of the great size of the ship. Coming up the ladder at the side doubtless gave her an idea of the height and then a walk from stem to stern an estimate of her length. It was touching to see the radiant smile on that poor girl's face as she listened to the boy, who told her of the wonders he saw.

For the first two weeks of the cable's operation, messages were carried by steamer between Newfoundland and Nova Scotia. On 13 August, the Heart's Content cable station received word that the cable across the Cabot Strait had been repaired and was in good working order. The *Bloodhound* had found the break in seventy fathoms of water a few miles from shore, and made the repairs. The repair was supervised by Alexander M. Mackay, who was then Superintendent of the New York, Newfoundland and London Telegraph Company.[50] A successful North America to Europe cable system was finally operational. However, a further problem was to arise. After direct communication to the mainland was reestablished, a storm came up on 18 August, knocking out the overland cable between Bay du Nord and Grandy's Brook on the south coast of the island of Newfoundland. Direct connection to Cape Breton Island was again out of service and was not restored until the overland cable was repaired on 26 August.

News of the successful transatlantic cable quickly spread throughout Europe and North America, and major celebrations were held in New York to commemorate the occasion. In St. John's there was widespread rejoicing and celebrating, the like of which the city had never seen. An account of the celebrations was given in the *St. John's Patriot*:

> There were public rejoicings on Friday and Saturday, August 3rd and 4th, throughout the city. There was a pyrotechnic display from the turrets of the Colonial Building, which was festooned with lamps and brilliantly illuminated. Fireworks and musketry added to the festivities.

UNFINISHED BUSINESS

Although it had completed its main task, the *Great Eastern's* work was not yet finished. It loaded on board the 600 miles of cable carried by the *Medway*, and took on 8,000 tons of coal, which had been previously shipped from Britain. On 9 August, the *Great Eastern* and the *Medway* headed east to retrieve and splice into the lost 1865 cable.[51]

The *Terrible* and the *Albany* were already at the site and had already begun grappling operations. On 13 August the *Great Eastern* joined in the search. The cable was easily found; however retrieving it was another matter, as it was under more than two miles of the Atlantic Ocean. With the technology of the time, it was astounding that it could be located, much less recovered. Throughout the month, thirty attempts were made to recover the cable. On several occasions, it was hoisted, but either it slipped away or the grappling rope broke. After a couple of weeks, the supply of grappling rope was running short and there was concern that the recovery operation would have to be abandoned. It was therefore decided to move the recovery operation about 80 miles to the east, where the water was not as deep. At the new location, the *Great Eastern* managed to grapple the cable and attach it to a buoy, which itself weighed more than three tons but was nevertheless capable of supporting more than four times its weight. It then moved several miles west and again grappled the cable. The *Medway* then moved two miles west of the *Great Eastern*, where it successfully grappled and cut the cable. This reduced the tension and enabled the *Great Eastern* to hoist the cable on board for testing. At this point, the ships were approximately 368 miles east of Heart's Content.

The Valentia telegraph office had been previously advised that the *Great Eastern* would be attempting to retrieve the 1865 cable and continuously monitored the cable for signals. On 2 September the station finally received a message from the *Great Eastern*. Valentia promptly telegraphed back that reception was perfect and that the cable was in good working order. Within a couple of hours, the *Great Eastern* spliced into the cable and headed back to Heart's Content, laying a new length along the way. News of the *Great Eastern's* success was wired to Heart's Content from Valentia and preparations for the landing of the second cable quickly began. On 8 September, the *Great Eastern* arrived at Heart's Content and finally landed the 1865 cable.

Celebrations again broke out after this success. Samuel Canning, who was in charge of retrieving the cable, was cheered by the crowd as well as by members of his crew as he came ashore. He was pushed into a chair that someone had brought along to view the arrival of the expedition and was carried on the backs of a group of men through the streets of the town. Within a period of six weeks, there were two fully operational cables between Europe and North America. The following day, after a productive mission, the *Great Eastern* fired four guns and left Heart's Content for Liverpool, arriving there on 19 September.

The cable as well as the earlier one was terminated in a house owned by Elias Warren, a local fishing and sealing merchant. The company had not made more permanent arrangements for a cable station pending the successful landing of the cables. Warren's house was only used for a couple of months until a new cable office could be constructed.[52] It was built by J. & J. Southcott, a St. John's architect and builder. Land for the building was purchased from Warren. The Heart's Content cable office log for 17 November, 1866 states "Cables removed to new offices in afternoon, work not interrupted."

On 12 September, the *Medway* and the *Terrible* left St. John's and headed for the Cabot Strait, where they placed a new cable to Cape Breton Island. At Cape Breton Island overland lines connected into Western Union's lines to the United States. Both transatlantic cables immediately went into commercial service. The initial transatlantic telegram rate was £20 for the first twenty words, including name, address, date and signature, and £1 for every word thereafter.[53] At such high rates, the service was primarily used for business purposes, and in the first two months of operation, almost 2800 messages were carried. As the technology and speed of transmission improved, the rate dropped. By the late 1890s it was down to one shilling per word.

Field's goal had finally been completed. After twelve years of frustration and failures, he had achieved the dream that occurred while he was studying his globe shortly after Gisborne's visit in 1854. The North American continent was now connected telegraphically with Great Britain, Europe and most of the world.

"The Atlantic telegraph cable of 1865: Chairing Sir Samuel Canning at Heart's Content, Newfoundland," from the London Illustrated News, 13 October 1866

Nine

EXPANSION OF TRANSATLANTIC TELEGRAPHY

After the Anglo-American Telegraph Company demonstrated the viability of transatlantic telegraph communications, other companies became interested in capitalizing on the commercial potential of the business. The most significant of these were the Direct United States Cable Company (Direct Cable), the Commercial Cable Company (Commercial Cable), Western Union Telegraph Company (Western Union), and several French and German concerns. The early transatlantic cable companies were primarily commercial ventures, but some also entered the business in order to meet their countries' national interests in providing their own transatlantic communications systems.

One of Anglo-American's early competitors was La Société du Cable Transatlantique Française (SCTF). This company was set up by Baron d'Erlanger and Paul Julius van Reuter. Reuter was one of the pioneers in news gathering and one of the first to use telegraph lines for this purpose. The current Reuters News Agency was founded by him. Although SCTF was organized in France, it was effectively under British financial control.[54] In 1869, the company installed a cable between Brest, France, via St. Pierre, a French colony off the south coast of Newfoundland to Duxbury, Massachusetts. The Brest – St. Pierre section was laid by the *Great Eastern* under the command of Captain Robert Halpin, who had been involved in laying the 1866 Heart's Content cable. Halpin would later marry Jessie Munn, from Harbour Grace, the daughter of a prominent merchant. The 750 mile section between St. Pierre and Duxbury was laid by the *Chiltern*.

The connection to Duxbury was completed on 23 July 1869, and its residents celebrated the completion of the new cable with the same enthusiasm that residents of Heart's Content celebrated the first successful transatlantic cable three years before. Ceremonies observing the cable completion were extensive and included a message to Napoleon III, the emperor of France.[55] SCTF did not last long as an enterprise, and merged with the Anglo-American Telegraph Company in 1873.

A few years later another French company entered the transatlantic cable business. The company, named La Compagnie Française du Télégraphe de Paris à New York, was popularly known as the "P.Q. Company," after Monsieur Pouyer-Quertier, one of its main proponents. In 1879, the P.Q. Company hired Siemens Brothers of Germany to install a cable between Brest and St. Pierre. This cable was later connected via Cape Breton to Cape Cod. In 1894, the P.Q. Company amalgamated with La Compagnie Française des Télégraphes Sous-marins and became La Compagnie Française des Cables Télégraphiques. [56]

In 1898, the new company installed a telegraph line between Brest and Cape Cod, as well as a connecting line to Coney Island.[57] The line was manufactured by La Société Générale des Téléphones and was 3,173 nautical miles long. It was laid by the *François Arago*, which installed the cable in four voyages, one section at a time. The cable was called "Le Direct," and at the time was the world's longest continuous underwater cable.[58]

In 1900, the German Atlantic Telegraph Company laid a cable from Borkum, Germany to New York via Fayal, Azores. The section between Germany and Fayal opened for service on 26 May, and the New York section opened on 28 August. In 1903, the company commissioned another transatlantic cable, which was installed by the *Stephan* along the earlier route. The Borkum – Fayal section was completed in 1903 and the New York section in the summer of 1904. The German Atlantic Telegraph Company used the cables until the beginning of World War I. After hostilities broke out, the British Navy cut the Borkum-Fayal section of the German cable and used it for its own purposes. In 1917, the eastern end of the 1900 cable was diverted by the *Colonia* to Penzance. Also in the same year the British diverted the 1903 Fayal – New York cable to Halifax. The British Post Office continued to operate the cables until 1929, when they were taken over by the Cable and Wireless Company. France took over the western end of the 1900 cable and diverted the eastern end of the 1903 cable to Brest.[59]

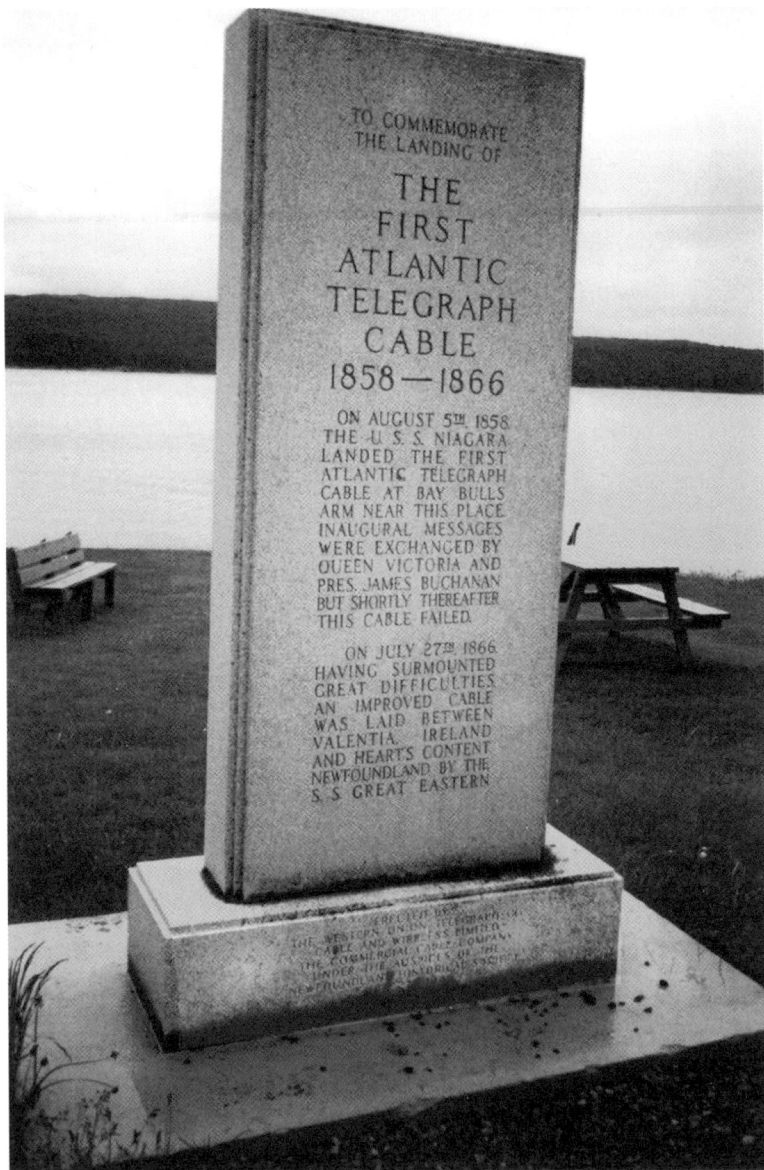

TO COMMEMORATE
THE LANDING OF

THE FIRST ATLANTIC TELEGRAPH CABLE
1858—1866

ON AUGUST 5TH 1858
THE U.S.S. NIAGARA
LANDED THE FIRST
ATLANTIC TELEGRAPH
CABLE AT BAY BULLS
ARM NEAR THIS PLACE
INAUGURAL MESSAGES
WERE EXCHANGED BY
QUEEN VICTORIA AND
PRES. JAMES BUCHANAN
BUT SHORTLY THEREAFTER
THIS CABLE FAILED.

ON JULY 27TH 1866
HAVING SURMOUNTED
GREAT DIFFICULTIES
AN IMPROVED CABLE
WAS LAID BETWEEN
VALENTIA, IRELAND
AND HEART'S CONTENT
NEWFOUNDLAND BY THE
S.S. GREAT EASTERN

Monument at Heart's Content to commemorate the 1858 and 1866 Atlantic cables.

Ten

ANGLO-AMERICAN/WESTERN UNION – HEART'S CONTENT

In 1866, Anglo-American amalgamated with the Atlantic Telegraph Company and obtained its monopoly on telegraph cables in Newfoundland. Atlantic Telegraph had earlier acquired the monopoly by merging with the New York, Newfoundland and London Telegraph Company.

After the Heart's Content 1865 and 1866 cables went into operation, transatlantic telegraph communication became popular on both sides of the Atlantic. It did not take long for businesspeople and ordinary citizens in Europe and North America to see the value in the new communications link and it became an instant success.

In the early days of transatlantic telegraphy, it was the custom for cable stations to log cable activity and operating conditions. One of the earliest surviving logs for Heart's Content was for 1 May 1867, in which the weather and number of messages handled were recorded: "Fine, fresh, forwarded 27 received 45." Despite the high rates for transatlantic messages, the log for the month of August 1867 shows the station received 809 messages and forwarded 725, providing revenues of about £21,000.

Anglo-American's cables connected at Heart's Content with the lines of the New York Telegraph. Both companies occupied the same building, with the two operations separated by a partition. Messages received by one company were passed on to the other for re-transmission. Anglo-American was known as the "English" company, while New York Telegraph was known as the "American" company.

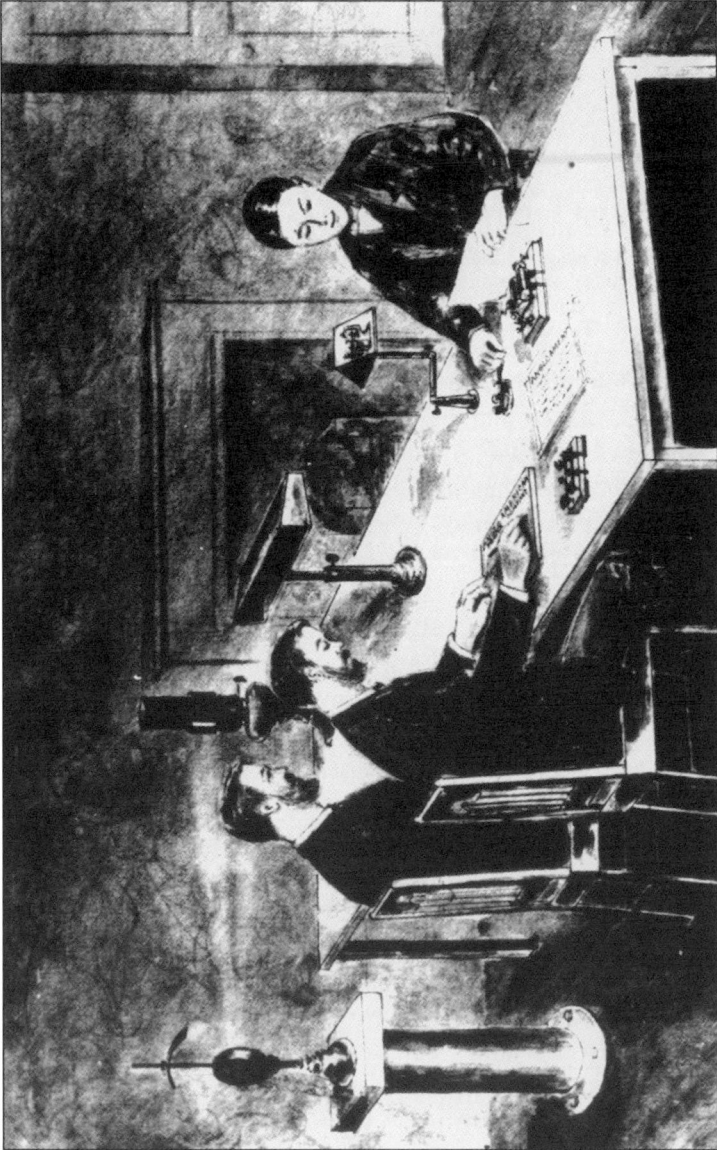

Heart's Content Operators (Courtesy of PANL)

At the time, Heart's Content connected into the Newfoundland telegraph system via an overland cable to Harbour Grace. From Harbour Grace the system connected to Brigus and to points east to St. John's, and to points west along the south coast of Newfoundland to Cape Ray, where an underwater cable crossed the Cabot Strait to Sydney.

Anglo-American's early employees were from Great Britain and Ireland, whereas some of New York Telegraph's employees were Newfoundlanders. In November 1866, after Collett and Lundy returned to England, Ezra Weedon became clerk in charge of Anglo-American. In March 1867, he was promoted to superintendent. John Waddell was manager of New York Telegraph's station, while Alexander Mackay was overall superintendent of their Newfoundland operations.

One of the Newfoundlanders who worked at the early Heart's Content station was Samuel Bailey. The station's records show that he was employed there as early as 1870. Recently some letters were discovered written by Bailey to his mother in 1867, while he was employed there by New York Telegraph. In a letter dated 9 November 1871, he mentions that Ezra Weedon had married a Miss Rutherford in Harbour Grace a fortnight earlier. After arriving by ship from Harbour Grace, the groom and his new bride were treated to a celebratory horse drawn carriage ride through Heart's Content.

As of April 1870, there were twenty employees with Anglo-American. The wives of five of the employees were living with them in Heart's Content and collectively had fifteen children. At the same time there were thirteen employees on the staff of New York Telegraph, of whom three had wives and a total of five children.

Anglo-American's employees at Heart's Content as of 3 April 1870 were as follows:

Isaac Angel
William Bellamy, wife, child, and servant
George Charlton
A. Courteen
James Bartlett, servant
Mr. Collins
Mr. Davis

William Dickenson, wife, 3 children, and 2 servants
J. W. James
Francis Perry
Mr. Peters
John Sullivan, wife, 5 children, and servant
Charles Trippe
George Unicume
Ezra Weedon (superintendent)
George Wickenden
William Woodcock, wife, child, and servant
Battery man, wife and 5 children
3 servants in the mess

New York Telegraph's employees in Heart's Content at the same time were:

Samuel Bailey
George Carson
Samuel Earle
Mr. Howell
H. H. MacKenzie
Mr. Mitchell
Edwin Moore
Alexander Saunders, wife, child, servant
John Smith, wife, servant
John Waddell (manager)
Battery man, wife and 4 children
2 servants in mess

A typical manpower shift at the Anglo-American office, recorded in the 24 April 1870 log, shows the following:

6 am-2 pm	Woodcock and James
8 am-10 pm	Perry or Trippe
10 am-6 pm	Sullivan and Bellamy
2 pm-11pm	Bartlett and Angel
11pm-6am	Unicume and Davis

According to Weedon's log book for the month of July 1870, 4767 messages were forwarded and 6368 received for a total of 11,135 messages. This level of activity produced revenues of approximately £29,000 to the company.[60]

In the early days of transatlantic submarine telegraphy, service was frequently interrupted because of ocean disturbances and cable breaks caused by icebergs and ship anchors. It was common for transatlantic cable companies to rely upon one another for help when their systems went out of service. For instance in 1870, Anglo-American used the French cable system via Placentia and St. Pierre when Western Union's lines failed west of Canso, Nova Scotia. Similarly, the French system used Anglo-American's lines when their system was inoperable.[61]

For the first few years, times were not easy for the staff at the station. In the log book for 18 January 1868, Weedon complains about illnesses caused by drafts. He wrote that Perry and Crocker were very ill and Charlton had been in bed for almost a month. The log book for 1870 mentions many sicknesses and notes that Woodcock's child had died, and that Bartlett's wife had died in St. John's.

Heart's Content was relatively remote. The only large town in the area was St. John's, which was a day's voyage away. To help compensate for the isolation, the staff at the station were fairly well paid. In 1869, the office superintendent received £300 per year, while the operators were paid between £150-200 per year.

The employees at the Anglo-American cable office in 1873 included Angel, Bartlett, Collins, Courteen, James, Newitt, Trippe, Unicume, Wickenden, and Williams. Employees involved with New York Telegraph's land line operation were Bailey, Collins, MacKenzie, Martin, Mitchell, O'Mara, Stentaford, and Thompson.

In 1873, Anglo-American Telegraph and New York, Newfoundland and London Telegraph merged into a single company and retained the Anglo-American name, making all employees members of that firm. Ezra Weedon was made superintendent of the entire operation.

1873 CABLE

Anglo-American installed its third cable from Valentia in 1873. The cable was originally ordered by SCTF for a line between Brest and Halifax. After gaining control of SCTF, however, Anglo-American decided to use the cable for the Valentia – Heart's Content route.[62] At Valentia on 14 June, the *Great Eastern*, under the command of Captain Halpin, spliced into the shore cable and commenced its voyage to Heart's Content. By that time, the laying of submarine cables was becoming routine, and the *Great Eastern* continued its journey across the Atlantic without incident. On 27 June, the *Great Eastern* buoyed the cable about 80 miles east of Heart's Content and proceeded to the town, arriving there in the afternoon. Accompanying the *Great Eastern* were the support vessels *Hibernia* and *Edinburgh*.

On 1 July, the *Hibernia* began paying out the shore end cable to connect with the transatlantic portion the *Great Eastern* had earlier buoyed offshore. By 5 July, the installation was completed and entire transatlantic cable was handed over to Anglo-American's technicians for a thirty-day test period.

Because the new cable was technically superior to the earlier ones, the project was important to Anglo-American. Cyrus Field travelled to Heart's Content from New York for the installation. Newfoundland's governor Stephen John Hill and his wife were also at Heart's Content to witness the installation, and were given a tour of the *Great Eastern* after a salute of seventeen guns.

By the time the 1873 cable became operational, the 1865 cable had failed beyond repair and the 1866 cable was working intermittently. The *Great Eastern* therefore headed back to sea and repaired the 1866 cable. This cable remained usable until 1877.[63]

The 1873 cable was terminated in the wood frame building constructed shortly after the 1866 installation. Anglo-American realized that a more substantial building was required and commissioned a permanent brick structure. The builders were once again J. & J. Southcott, who also constructed several residences for the staff of the station. Later, other staff buildings were erected, some of which are still being used as private residences.

After completion of the new cable office in 1876, the old wood frame building was given to the employees as a recreation complex and named Variety Hall. In 1919, Western Union used the hall as a training school where employees such as George Mallam, J. Allan Rowe, and T. Gordon Wilcox taught young men and women different aspects of telegraph communications.[64] The hall was the site of many meetings and entertainment events, but was difficult to maintain. It was therefore sold in 1955 and eventually demolished. Also in 1919, the cable station was enlarged to its present size by Saunders and Howell of Carbonear, who built additional residences for the employees. This building is now used for government offices as well as a museum, in which the original telegraph equipment can be seen.

With its 1873 cable, Anglo-American's intent was to improve the reliability of telegraph service between Great Britain and the United States. Although the new transatlantic cable performed flawlessly, the connection to the North American continent was unsatisfactory because of the poor connection to the mainland. As the overland line to Cape Ray was not in good condition, the company had to rely on its cable to Cape Breton via Placentia. It was therefore decided that a new cable to the mainland would be required. Rather than build another overland route, Anglo-American decided on a new submarine facility to Nova Scotia. In November the *Robert Lowe* installed a cable from Heart's Content to Rantem, Trinity Bay. From Rantem, a cable was trenched across the isthmus of Avalon to Island Cove, Placentia Bay, where it connected to a new submarine cable (via Placentia) to Sydney, Nova Scotia. A new cable was also laid between Placentia, St. Pierre and Sydney. This cable was used to connect to the mainland as well as with SCTF's system at St. Pierre, which Anglo-American took over in the same year. The

The Heart's Content station as it stands today, now used as a museum

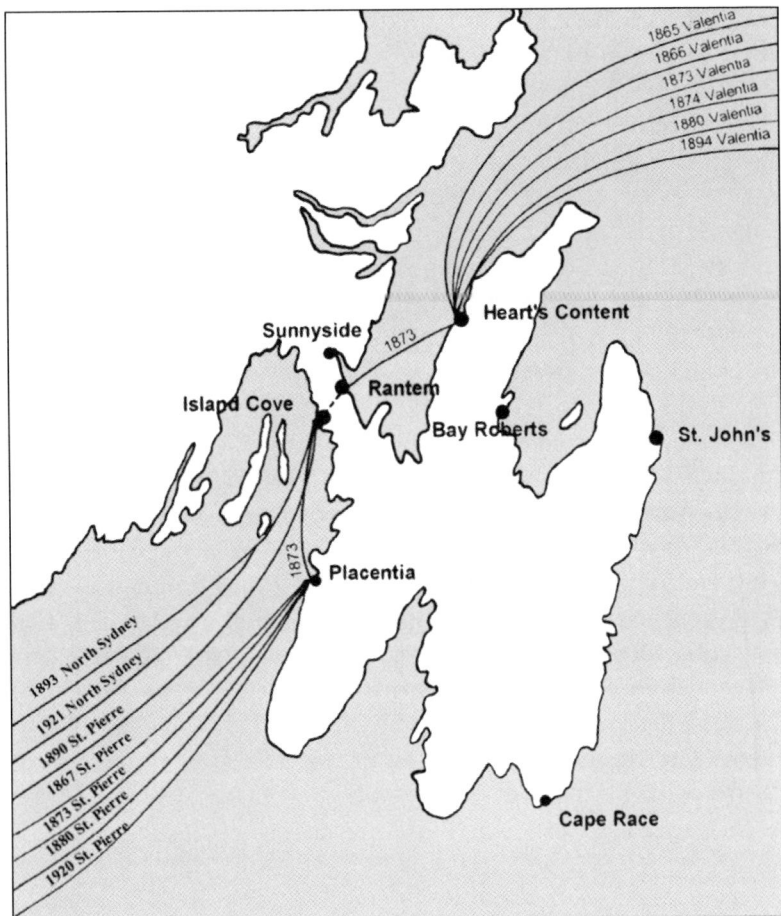

Some of the major cable landings in Newfoundland associated with the Heart's Content cable station

new routes provided additional capacity and solved the outage problems that plagued the Cape Ray line. Also following the Heart's Content – Island Cove route, additional cables to North Sydney were laid in 1893 and 1921, and to St. Pierre in 1890.

THE SINKING OF THE *ROBERT LOWE*

On 21 November, shortly after the installation of the 1873 cable, the *Robert Lowe* sank in St. Mary's Bay, with the loss of Captain Tidmarsh, sixteen crew and one passenger. The Heart's Content records show that on 26 November Samuel Bailey and Mrs. Tidmarsh left for St. John's. Presumably Mrs. Tidmarsh was

Captain Tidmarsh's wife. The sole passenger casualty was George Wickenden, an electrician at the Heart's Content station. A plaque to his memory was erected in St. Mary's church in Heart's Content.

1874 CABLE

In 1874, Anglo-American installed another link between Ireland and Heart's Content. It was again laid by the *Great Eastern*, which took advantage of the prevailing winds and installed the cable from west to east. The *Great Eastern* began paying out cable from Heart's Content on 26 August and arrived in Valentia two weeks later. The cable was quickly put into service and worked well for several years. In November 1881, it broke during a storm off the Irish coast, and was out of service until February 1882 when the *Gamecock* and the *Kangaroo* completed its repair.

1880 CABLE

Because of the growing demand for transatlantic telegraph service, Anglo-American installed another cable between Heart's Content and Valentia in 1880. This cable was essentially a rebuild of the one placed in 1866. The shore ends of the 1866 cable were still in good shape and could be reused; however, the entire deep water portion had to be replaced. The cable was laid by the *Seine*, under the command of Captain Halpin. The *Seine* left Heart's Content on 6 August and arrived at Valentia on 21 August. The rebuilt cable was now referred to as the 1880 cable and remained in service as late as 1949.[65]

According to the station diary, there were nine Anglo-American owned houses in 1881, and six scheduled to be built the following year. Messrs. Angel, Williams, Saunders, Bailey, Mackenzie and Stentaford were scheduled to occupy these houses. Also in 1881, there were nine single men living in the staff house. The staff included a housekeeper, housemaid and a kitchen maid. The staff house also contained a billiard room and library, where abstinence and an 11 pm curfew were rigorously enforced.

In 1883, there were forty-four employees at the Heart's Content station. It is not known why the station manager kept track of the employees by religion, but the station's diary shows that thirty-six were Church of England, five were Roman

Rusty remains of transatlantic cables at Heart's Content in 1998

Catholic, and three were Wesleyan. Anglo-American also regularly hired students, both male and female. In 1885, the staff at Heart's Content included fourteen operators, and ten employees involved with the land end of the operation.

Heart's Content staff houses circa 1890
(Courtesy of Heart's Content museum)

1894 CABLE

Anglo-American's last transatlantic cable was installed by Telcon in 1894. The Heart's Content end was laid by the *Britannia*,[66] and the transatlantic section by the *Scotia*. The *Scotia's* cable was more than 2000 miles long and weighed more than 5000 tons. On its way from Ireland, the *Scotia* encountered heavy fog off the Newfoundland coast and had to proceed at reduced speed. In the fog, it hit an iceberg, but because it was proceeding slowly, the ship incurred little damage. Nevertheless, the collision was serious enough to order the lowering of lifeboats, and one member of the crew dropped dead from a heart ailment. The ship's cable-laying equipment was slightly damaged, but repairs were later made at Heart's Content, which the *Scotia* reached on 10 July. The *Loughrigg Holme*, carrying a supply of coal, had already arrived and had fuelled the *Britannia*. The coaling of the *Scotia* was completed on 13 July. Meanwhile, repairs to the *Scotia's* cable-laying equipment proceeded and were completed on 14 July. The *Britannia* laid 237 nautical miles of cable out from Heart's Content, and spliced it into the cable laid by the *Scotia*. The total length was 1847 nautical miles and the installation was completed by 27 July.

The message-carrying capability of the new installation was ten times greater than the 1866 cable and greatly improved communications between the two continents. In addition to personal, political and commercial news, much of the traffic carried over the cable was information regarding the stock markets on both sides of the Atlantic.

PLACENTIA CABLE STATION

In the early days of transatlantic cables, Placentia played a significant role in relaying messages between Heart's Content and the mainland. The earliest cable was placed between Placentia and St. Pierre in 1867.[67] The cable was owned by New York Telegraph, and the Placentia manager at the time was a Mr. Roche.

In 1873, Anglo-American installed a Placentia – Sydney cable via St. Pierre. There were also land lines between Placentia and Heart's Content, which were used for transatlantic traffic over the years. In the early days, Placentia served as a repeating station for duplex circuits. (Duplex technology allowed messages to be both transmitted and received over a cable at the same time, effec-

The old Placentia telegraph station (Courtesy of Newfoundland Historic Trust

tively doubling its call-carrying capacity.) In 1878, the repeating equipment was removed and the cables were operated directly between Heart's Content and Sydney. In 1880, the *Seine* laid a three-core cable between Placentia and St. Pierre which was later extended to North Sydney by the *Kangaroo*. In 1895, there were two cables working between Placentia and St. Pierre and two from Placentia to Sydney.[68] As previously mentioned, there was also a cable to Island Cove, Placentia Bay, where the cable was trenched and connected to the Rantem – Heart's Content cable.

Another cable was laid between Placentia and St. Pierre, in 1920. In the following year, a cable was installed between Island Cove, Placentia Bay and North Sydney, which connected to Canso in 1922.[69] The 1873 Anglo-American telegraph office was located in the old Sweetman residence, which is one of the oldest buildings in Placentia.

THE END OF ANGLO-AMERICAN'S MONOPOLY

Anglo-American's monopoly expired in 1904. This presented an opportunity for other companies to locate in Newfoundland and take advantage of the shorter underwater distance for their transatlantic cables. The shorter submarine cable enabled messages to be transmitted at a higher speed. The resulting increase in the number of messages carried reduced the cable companies' cost and hence increased their profits. The termination of Anglo-American's monopoly opened up the market to competition, and shortly after, Commercial Cable set up in St. John's. Direct Cable also established a station in Harbour Grace, as did Western Union in Bay Roberts, but these two stations – along with the Heart's Content station – soon came under the control of Western Union. More on these stations is contained in the following chapters.

Anglo-American leased its transatlantic cables to Western Union for a 99-year period in 1912.[70] This long-term lease effectively made Western Union the operator of Anglo-American's cables. The Anglo-American name continued to be associated with the Heart's Content station for some time, but the Western Union name eventually took over.

In 1914, the staff of the Heart's Content cable station was approximately 80. During World War I, the volume of telegraph traffic rose dramatically and the staff increased to almost 300 employees, including 60 women. During both world wars, Heart's Content was a critical communications station between North America and Europe. During World War II, the station was patrolled by members of the Newfoundland constabulary, which provided 24-hour protection.

C. H. Tranfield, superintendent of the Heart's Content station, wrote a letter to his superiors at Western Union in New York in 1918, stating that there was trouble on the land line to Placentia because youths were breaking the insulators. Also in 1918, he noted that there were twenty-five operators, one housekeeper, four maids and one janitor at the Heart's Content staff house.

Anglo-American's employees were very well paid relative to other residents of Heart's Content. The 12 October 1918 diary lists the following employees and their wages:

Employee	Title	Monthly Salary
C. H. Tranfield	Superintendent	$260.00
J. F. Richards	Asst. Superintendent	207.03
F. Anderson	Supervisor	222.99
G. C. Bailey	Multiplex Supervisor	129.39
G. Carberry	Supervisor	199.79

R. B. Comerford	Multiplex Supervisor	187.72
A. Farnham	Supervisor	201.95
W. N. Ford	Supervisor	216.26
E. Mallam	Supervisor	174.85
M. M. Hopkins	Supervisor	194.21
W. Mallam	Supervisor	198.83
W. R. Moore	Supervisor	145.12
W. C. Palmer	Supervisor	198.64
M. A. Rabbitts	Supervisor	178.85
C. R. Rowe	Supervisor	201.42
R. T. Tobin	Supervisor	189.67
James Wilcox	Supervisor	191.53
Total		$3298.25

In the station's weekly report of 14 October 1918, the following statistics were recorded:

Total number of words handled	1,205,870
Total operator hours used	1,823
Average per operator hour	661
Number of hours overtime	1,025
Number of checker's hours	904

Some of the Heart's Content cable station staff circa mid 1950s. Front row: Dr. B. J. Short, John Stentaford, Evan Pugh (superintendent), Andrew Hillyard, Archibald Budden. Back row: George Ashley, John Bonfield, Willis White. (Courtesy of Mr. Robert Balsom)

Old telegraph equipment on display at the Heart's Content cable museum

Opposite page: – *Staff of the Heart's Content cable station circa 1908, listed by number and column, from left to right, and top to bottom. Column one: 42. J. A. Rowe, 40. F. T. Peach, 46. John Farnham, 38. H. R. Rendell, 50. James Legge; Column two: 44. A. Collins, 24. F. Anderson, 22. J. C. Bailey, 20. W. Moore, 26. William Mallam, 47. L. James. Column three:28. A. George, 10. M. A. Rabbitts, 16. W. T. Stentaford, 18. G. C. Bailey, 30. James Farnham. Column four: 32. E. Mallam, 8. J. J. Sinnott, 2. J. W. James, 6. J. Scotland, 12. A. Farnham, 34. G. Carberry. Column five:14. M. M. Hopkins, 4. J. F. Richards, 1. William Bellamy, superintendent, 5. S. S. Stentaford, 15. J. Ollerhead, 36 R. T. Tobin. Column six: 33. J. G. Joy, 9. J. Wilcox, 3. C. H. Transfield, 7. Dr. Anderson, 13. J. Cook, 35. A. R. Martin. Column seven:29. George Butt, 11. W. N. Ford, 17. J. A. Scotland, 19. R. B. Commerford, 31. G. S. Mallam. Column eight: 45. R. J. Bemister, 25. W. C. Palmer, 23. G. S. Butt, 21. T. J. Oates, 27. W. Pugh, 48. C. C. Butt. Column nine: 43. C. Hartley, 41. D. H. Cameron, 37. C. R. Rowe, 39. H. R. Rendell, 49. T. G. Wilcox..*

Staff of Heart's Content Cable Station in 1908
(Courtesy of Heart's Content museum)

FEMALE EMPLOYEES

The 1 August 1919 monthly report of the Heart's Content cable station stated that "as soon as possible, twelve additional female students will be engaged." This high level of staffing by women was probably a landmark in the history of female employment in Newfoundland. In fact in 1919, the Heart's Content station employed at least 42 women, as listed below:

Miss Williams	Miss Rabbitts
Miss F. L. Rowe	Miss King
Miss Oates	Miss N. Mallam
Miss K. F. Richards	Miss Margaret Rowe
Miss E. Earle	Miss C. Duff
Miss N. Reid	Miss Pinsent
Miss L. V. Mitchell	Miss B. Bartlett
Miss Howell	Miss B. Matthews
Miss Leslie	Miss C. Baggs
Miss Aitken	Miss A.S. Mitchell
Miss E. Ollerhead	Miss O. Farnham
Miss M. J. Wilcox	Miss E. Mitchell
Miss Bessie Matthews	Miss L. Sinyard
Miss A. Moore	Miss Lever
Miss F. K. Richards	Miss V. Pugh
Miss Allison Rowe	Miss A. Randell
Miss F. Roil	Miss A. Bartlett
Miss D. Moore	Miss Williams
Miss Mallam	Miss Dora Moore
Miss Jessie Dawe	Miss Margaret Rowe
Miss May Smith	Miss J. L. Legge

These names were garnered from the absence and sickness records of the cable station and are therefore not complete. The full names are not available for the above, neither are the particular jobs or functions that each person held. In all probability, their jobs were clerical in nature, such as recording and checking messages; however, it is also likely that many of these female employees were also operators. The records indicate that they were all "Miss," which is probably more a commentary on the work practices of the time rather than Anglo-American's hiring practices.

ANGLO-AMERICAN'S ST. JOHN'S OFFICE

In 1919, superintendent Tranfield wrote to Western Union in New York advising them that work on renovating the Anglo-American telegraph office (used for local telegraphs) in St. John's was progressing favourably. He also ordered a counter and a gate, as well as floor tiles to be used to imprint the initials AATC standing for Anglo-American Telegraph Company. This office was located in the vicinity of the present-day Scotia Tower on Water Street. In September 1938, by which time teleprinters were in service, Anglo-American inaugurated a direct circuit to from St. John's to Western Union's office in New York City. This circuit was inaugurated by St. John's mayor Andrew Carnell and J. M. Barbour, Anglo-American's superintendent in Newfoundland.

TRANSMISSION OF TELEGRAPHS

Up to the 1920s, overseas messages received at Heart's Content, as well as at the cable stations in St. John's, Bay Roberts and Harbour Grace, were manually re-transmitted by the operators. The signals were sent using a double key. Pressing the right key would send a positive voltage over the cable, representing a "dot." Pressing the left key would send a negative voltage, representing a "dash." For the first few years of transatlantic telegraphy, a mirror galvanometer was used at the receiving end to decipher the message. In the early 1870s, this device was replaced with a siphon galvanometer. The siphon galvanometer operated on the same principles as its predecessor but instead of using a mirror it employed a thin glass tube which siphoned ink out of an inkwell to make a mark on a moving strip of paper. When no message was being sent, the device would draw a line down the centre of the paper strip. When a positive signal (dot) was transmitted, a mark would be made above the line. A negative signal (dash) would produce a mark below the line. This produced a hard copy of the message, which greatly improved message accuracy.

After automatic repeating equipment became available in the 1920s, the need for operators was eliminated, except those required for maintenance purposes. This caused the staff level at Heart's Content to drop dramatically from almost 300 employees to approximately 30. Most of the employees affected were transferred to other Western Union locations in the United States and elsewhere.

Automatic repeating equipment received the overseas dots and dashes, and amplified and cleaned up the signal so that it could be regenerated and re-transmitted to its next destination. The amplifying and rotary regenerating equipment which performed this operation is currently on display in the Heart's Content museum. In the mid 1900s, the transmission speed over the transatlantic cables was approximately 300 letters per minute. During the 1950s, submarine repeaters (or amplifying devices which boost the signal and improve cable performance) were installed in Heart's Content's transatlantic cables several hundred miles from the station. Multiplexing equipment was later added, providing six separate channels per cable. This new technology greatly increased the calling capacity of the cables.

An important part of the Heart's Content station was the "artificial line" room. This room housed sensitive equipment to "balance" the submarine cables; this enabled duplex operation (i.e., simultaneous two-way communication) to work properly. The room was temperature controlled, with large ceiling fans and sealed doors. Currently, it is used as a projection and meeting room in the Heart's Content museum. After submarine repeaters were installed for the transatlantic cables, balancing was not required, but the artificial line room was still used to balance those cables going to the rest of North America.

Whenever cables went out of service because of breakage along the ocean floor, sensitive equipment (consisting of a Wheatstone bridge, resistance bank, and galvanometer) was used to accurately determine the distance from shore where a fault had occurred. A cable repair ship was then dispatched to grapple the cable, pull it up on deck and make the necessary repairs.

Electric power for the Heart's Content station was initially provided by copper-zinc dry-cell batteries. Later, when electric generating equipment became available, it became more efficient to use lead-acid storage batteries, which required charging. The station's batteries were located in the basement of the cable station. Electric power for the station was provided by a commercial hydro-generating station which was built in Heart's Content in the early 1920s. Early in the 1950s, a rectifier system was installed to convert the commercial power alternating current to the direct current required for the operation of the cable. Once this was completed, the batteries became redundant and were removed. A "no-break" diesel system was installed, ensuring that service would not be interrupted in the event of a commercial power failure.

THE CLOSING OF THE OFFICE

With advances in technology, the Heart's Content station became obsolete, and closed down in 1965, almost one hundred years after the completion of the first successful cable. Some of the employees of the Heart's Content station at its closing were B. C. Berrigan, F. A. Cumby, R. Cumby, C. George, G. C. Green, E. Harnum, C. Hobbs, G. W. James, M. N. Parrott, D. S. Pike, A. C. Tavenor, K.L. Traverse, and R. S. Stentaford.[71] Most employees were relocated to other Western Union offices such as New York and San Francisco. The closure of the Heart's Content cable station brought the town's exciting history of transatlantic cables to an end.

"The Atlantic Cable - View of Heart's Content, Newfoundland," from Harper's Weekly, *11 August, 1866*

Heart's Content cable station staff circa late 1920s. Front row: George Bailey, Ed Hopkins, Raymond Hopkins. Second row: William Moore, Moses Rowe, Stephen Hobbs. Third row: Harold Martin, Evan Pugh, Thomas R. Hopkins, John J. Young, T. Gordon Wilcox, Maxwell Young, J. Allen Rowe. Fourth row: J. W. Bonfield, Edwin Mallam, W. N. Ford, Roger Tobin, Henry Rendell. Back row: G. A. Young, H. Ernest Wyatt. (Courtesy of Mr. Claude Hobbs)

Eleven

WESTERN UNION TELEGRAPH COMPANY – BAY ROBERTS

The Western Union Telegraph Company (Western Union) was set up in New York in 1851 to construct a telegraph line between St. Louis and Buffalo, New York. By the early 1860s, the company had built extensive telegraph lines throughout the United States.

In 1881, Jay Gould (1836-1892), an American tycoon and major shareholder of Western Union, set up the American Telegraph and Cable Company (American Telegraph), which also had an interest in the transatlantic telegraph business. The company installed a cable between Penzance in Cornwall, Great Britain and Canso, Nova Scotia, and then leased it to Western Union in 1882 for a fifty-year term. The cable was contracted to Siemens Brothers (Siemens) and was installed by the cable ship *Faraday*.

Gould was noted for his questionable business practices. He purchased shares in Anglo-American, a competitor of his company, and then spread word that American Telegraph's cable had failed. He went as far as to keep his company's cable out of service to help solidify this rumour. As a result, the value of Anglo-American's shares increased. Gould sold his stock in Anglo-American, and with the profits commissioned a new cable for American Telegraph. In the following year, American Telegraph contracted with Siemens to lay a second transatlantic cable which followed the same route as the earlier one.[72] In 1889, both cables were extended to New York.[73] Some time later, Western Union purchased American Telegraph for $2,000,000.[74]

A 1920 view from the bay with Cable Avenue to the left and the Bay Roberts cable building on the right (Courtesy of Mr. Raymond Norman)

In 1912, Western Union entered into ninety-nine year operating leases on the transatlantic cables owned by Anglo-American and Direct Cable. This effectively gave Western Union control of those companies, although they continued to operate under their own names. Western Union's lease arrangements meant that its only remaining competition was the Commercial Cable Company, which will be discussed later.

At the end of Anglo-American's monopoly, Western Union decided to locate in Newfoundland. In 1910, it set up a cable station at Bay Roberts and contracted with Telcon for the installation of a cable between Hammel, New York and Penzance.[75] The cable was routed via Bay Roberts and laid by the *Colonia* and *Teleconia*. It was sold to Anglo-American just before Western Union and Anglo-American entered into their 1912 operating agreement.

In 1913 and 1915, respectively, the company diverted its 1881 and 1882 Penzance cables to Bay Roberts. By diverting these cables, the speed of telegraph transmission was greatly improved. The Canso end of the cables was discarded and in 1913, the company installed a new cable from North Sydney to Bay Roberts. The cable came ashore at Colinet, St. Mary's Bay, and

Cableships at Bay Roberts in 1926. The Teleconia *is in the foreground and the* Colonia *is in the background (Courtesy of Mr. Raymond Norman)*

was terminated in a small building, where the undersea cable was spliced into the overland cables and trenched to Bay Roberts.

In the following year, Bay Roberts temporarily became the terminus for the Harbour Grace cable station when the two stations were linked via an underground cable.[76] This arrangement remained in place until the Harbour Grace operation was acquired by Cable and Wireless in 1920.

The 13 October 1916 edition of the *Bay Roberts Guardian* lists the following staff at the Bay Roberts cable station who each contributed $1.00 to the Women's Patriotic Association, an association of women who raised money for Newfoundland's overseas forces during World War I.

G. Ashley	J. Kielly
C. Bailey	G. F. Mackey
F. Bateman	B. Mercer
W. T. Bellamy	H. Noseworthy
R. Bemister	P. O'Leary
J. Gordon	H. Payn
J. Hambling	F. Peach
A. Howard	W. Pugh
L. Hurst	

During the 1920s, changes in technology, particularly the introduction of loaded cables, increased the speed at which messages could be sent. Loaded cables used special electrical devices

Landing the 1926 cable at Bay Roberts (Courtesy of Mr. Jack Hambling)

installed along their length to provide improved electrical performance. The early loaded cables operated at a high transmission speed; however, unlike unloaded cables, they could only handle messages in simplex mode (only one direction at a time). In 1924, Western Union installed a loaded cable from New York to Horta in the Azores, which had a speed of 1500 letters per minute.[77] At Horta, Western Union exchanged cable traffic with German and Italian cable companies, which carried the telegraph messages to Europe. On 8 July 1926, a new loaded cable from Penzance to Bay Roberts was laid by the *Colonia*. This new installation provided an even greater speed of 2400 letters per minute. The circuit was divided into eight channels, each operating at 300 words per minute. Before proceeding on to New York, the crew of the ship organized a dance, to which the staff of the Bay Roberts station was invited.

On 22 August 1926 the *Colonia* continued its installation and completed the cable to Hammel, New York.[78] In 1928, after duplex capability was developed for loaded cables, the same ship installed a duplex cable between Bay Roberts and Horta.[79] At the time, the 1928 cable was the fastest cable in the world, capable of speeds of 2800 letters per minute and of carrying four messages simultaneously in both directions.[80]

Western Union was to introduce a major advance in 1950 when it installed a submerged repeater in its 1881 cable.[81] This was the first use of a submerged repeater in a transatlantic cable, boosting the level of the overseas signal and greatly improving the cable's performance.

The first Bay Roberts cable station was a wooden structure on Water Street. The building still exists and is now used as a private residence. The structure served the company until a large brick building was completed on Water Street during World War I. Located in the new building were offices, an operating room, test room, artificial line room, maintenance shop, battery room, and electric generating plant. The grounds of the cable station were kept in excellent condition, and the building was surrounded with a beautiful lawn. The main operating room of the station had three wall-mounted clocks, under each of which was the slogan "Accuracy First." In 1998 the building was renovated, and is now occupied by the Bay Roberts town council.

The company built a large staff house and attractive homes for its senior staff on Cable Avenue, immediately west of the cable building. The houses had their own water and sewer systems, garbage collection, and firefighting protection.[82] The area was fenced, with entry to the avenue through one gate for vehicles

Cable Avenue, Bay Roberts, with the staff house in the background
(Courtesy of Mr. Raymond Norman)

and another for pedestrians. Included in the Cable Avenue complex was a cricket field, icehouse, two tennis courts, and lighted sidewalks, making this area of Bay Roberts a very elite district at the time.

The Bay Roberts station was connected to the station at Heart's Content via six aerial telegraph lines. As both stations were under control of Western Union, they cooperated with one another in time of difficulty. It was common, for instance, for one station to route some of its telegraph traffic via the other if its facilities were busy or out of order.

WORLD WAR II

During World War II, security at the Bay Roberts and other Newfoundland cable stations was very tight. When the Battle of the Atlantic was at its height, wireless communication was not used for the transmission of confidential information such as submarine and ship locations, and weather conditions along the Atlantic coast. This made Newfoundland's cable stations essential links in transatlantic communications. The Bay Roberts cable office complex was surrounded with barbed wire and was first protected by the Newfoundland Constabulary, later replaced by the Newfoundland Militia. Protective duty was eventually assigned to a detail of Canadian soldiers from Quebec, who with their bayonet rifles provided round the clock security for the facility.

More than thirty people were employed at the cable station at the height of its activity. By 1957, the number of staff was fewer than twenty. The last superintendent of the Bay Roberts cable station was Chester Smith, who was in charge of the operation when it closed down in 1960.

The staff of the Bay Roberts cable station circa 1957. Front row: Mr. C. G. Hopkins, Joseph D. Williams, Thomas A. Brien (superintendent), Chester E. Smith, Jack F. Hambling. Second row: Wallace Farnham, Robert J. Mercer, Noel Howard, Jim Burke, Ted F. Starr, Lou A. O'Brien, W. Spicer, Fred Starr, Harry Dawe. Back row: Fred Badcock, Jack French, Wiff Mercer, Fred D. Oates, H. Walsh. (Courtesy of Mr. Jack Hambling)

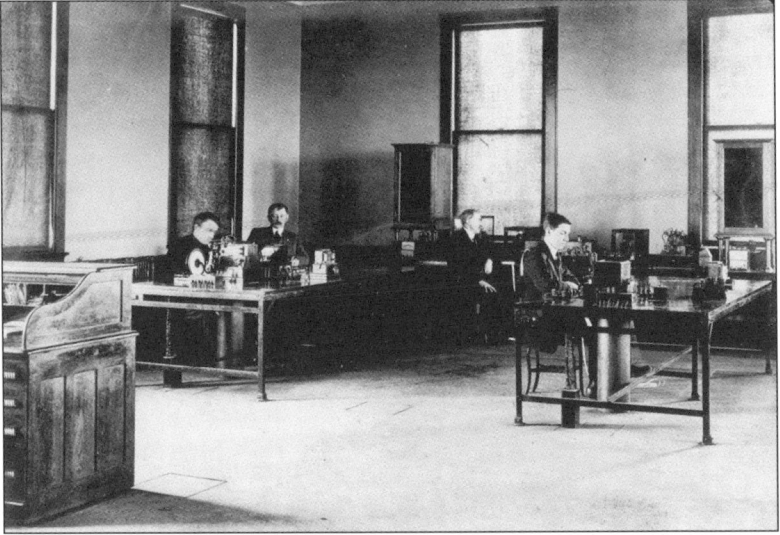

Operators in the Bay Roberts station circa 1920s (Courtesy of Mr. Jack Hambling)

The photographs above show some of the equipment in the Bay Roberts station circa 1920s (Courtesy of Mr. Jack Hambling)

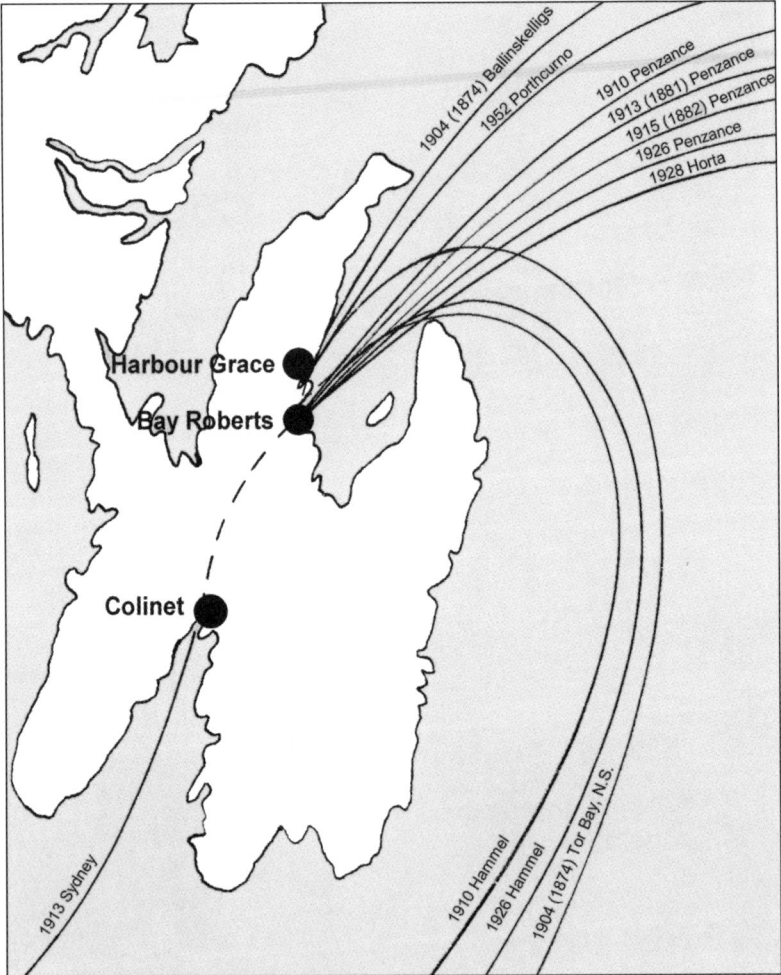

Major cable landings at Bay Roberts and Harbour Grace

Twelve

DIRECT UNITED STATES CABLE COMPANY – HARBOUR GRACE

The Direct United States Cable Company (Direct Cable) was formed in 1873 by John Pender to compete with Anglo-American for the transatlantic telegraph business. Direct Cable originally intended to lay a cable directly between the United States and Europe without landing in Newfoundland, Nova Scotia, St. Pierre or the Azores. It was thought that a system utilizing the shorter length of a direct cable would be less expensive than one landing at an intermediate location. After much investigation, it was decided that in spite of the shorter distance, a direct cable would have an adverse impact on the speed of transmission, and therefore a direct route would not be feasible.[83]

The following year, the cable ship *Faraday*, under contract to Siemens, began the installation of Direct Cable's first transatlantic cable. Because the ship could not carry all the cable in one load, it was laid in two sections. The first section extended from Rye Beach, United States, to Tor Bay, Nova Scotia and then to the head of Trinity Bay, Newfoundland, where the cable was attached to a buoy. After laying this section, the ship returned to Woolwich, England for more cable to complete the line from Trinity Bay to Ballinskelligs, Ireland.[84]

While the *Faraday* sailed to England, Direct Cable looked into the possibility of landing its cable in Newfoundland. The company pursued this option even though Anglo-American still held its monopoly on telegraph cable landings in the country. Having a station in Newfoundland would provide immediate benefits because it would shorten the transatlantic underwater length and improve the cable's transmission speed. Although Direct Cable challenged Anglo-American's monopoly, the monopoly was upheld by the Newfoundland court. After the company was denied landing rights, the *Faraday* spliced into the cable it had earlier buoyed in Trinity Bay, and continued the installation across the Atlantic.

Direct Cable's new transatlantic telegraph cable was very popular with its customers and the upstart company was doing well financially, earning 11% on its capital compared to Anglo-American's return of 7%. Anglo-American was threatened by Direct Cable's success, and John Pender, who was also Anglo-American's chairman, led a campaign to amalgamate the two companies, which he successfully achieved in 1877. Direct Cable, however, continued to operate under its own name.[85]

The Harbour Grace cable station as it lies derelict in 1998

After Anglo-American's monopoly expired in 1904, Direct Cable decided to take advantage of the shorter transatlantic distance by transferring its western terminus from Halifax , Nova Scotia (to which the cable had previously been diverted from Tor Bay) to Harbour Grace, Newfoundland. This was done in August 1910, when the company set up an office at Harbour Grace and diverted the 1874 Ballinskelligs cable to the new station. The cable diversion was completed by the *Colonia*. By terminating the western end of its cable in Newfoundland rather than in Nova Scotia, the company doubled the speed of telegraph transmission, allowing twice the number of messages to be carried.

The Harbour Grace station was located at Ridley Hall, a two and a half storey brick and stone structure that was built in the early 1830s. The building originally had been built as a residence for Thomas Ridley, a prominent merchant of the time. On 2 August 1866, his son Harrison held a gala party celebrating the success of the Heart's Content 1866 cable, which had been completed a few days before.

The cable offices were located on the first floor while the manager's residence was upstairs. The Harbour Grace station operated only one cable and was staffed by approximately a dozen employees, including a manager and three engineers, who rotated from their English offices on a three-year basis. The manager of the company in 1910 was Mr. T. Gothorpe.

THE PREEMPTION ISSUE

The charter of the New York, Newfoundland and London Telegraph Company (which later merged with Anglo-American) had given it a fifty-year monopoly on telegraph communications in Newfoundland. It had also given the government the right after twenty years to cancel the monopoly and take over all telegraph communications in the country. The charter outlined a process to arrive at a fair price that the government would pay the company if it decided to cancel the monopoly and take over its telegraph assets.

The government's twenty-year preemption option was up in 1874. During the previous two years, the government debated whether to exercise its option, and had great difficulty deciding if it should take over Anglo-American's monopoly. Anglo-American's competitors took advantage of the government's indecision and manipulated the stock market value of Direct Cable's shares by playing up the notion that the government was going to cancel the monopoly and take over the telegraph system. This rumour lowered the value of Anglo-American's stock and increased the stock price of its competitors. One of the leaders of this stock manipulation was Henry Labouchere, a British MP, who was also a major shareholder in Direct Cable. By playing up the possibility that the Newfoundland Government would pre-empt Anglo-American's monopoly, speculators were led to believe that Direct Cable would benefit if it could land its cables in Newfoundland. The price of the company's stock therefore soared on the stock markets.

Labouchere tried to persuade the government to pre-empt Anglo-American's monopoly in the following letter to Governor Sir S. Hill K.C.M.G.:

> "Should you, Sir, terminate the existing monopoly in 1874, and take over the land lines of the island, for the value of their plant and material, I am informed by responsible capitalists that they will be ready, if it be wished, to take over your lines at a rental, agreeing to lower the tariff, and to allow all cables to land on your shores, and to advance money on the guarantee of the rental to enable your colony to pay off Mr. Field. In fact, they are ready to enter into any arrangement with you which may facilitate the operation, so anxious are all commercial houses and our daily press to reduce the heavy cost of transatlantic telegrams.
>
> I have the honour to be, Sir,
> Your Excellency's most obedient humble servant,
> Henry Labouchere"

After long debate, the Newfoundland government decided to pre-empt Anglo-American's monopoly and approached the British government for financial assistance to purchase their assets. The British rejected this request so the Newfoundland Government was forced to reverse its position and had to cancel its plans to pre-empt. This left Anglo-American's monopoly intact for a further thirty years.

When the government announced its final decision on preemption, Direct Cable's stock plummeted – not however not before stock manipulators had realized a handsome gain of approximately £400,000.[86]

ALL RED ROUTE
Around 1914, Western Union consolidated its Conception Bay operations and connected the Harbour Grace station to the Bay Robert's station via an underground cable. In the same year,

the Newfoundland government contracted with Commercial Cable for its off-island telegraph business. This created a problem for the British government because Western Union and Commercial Cable were American owned and the Newfoundland government's action broke a link in the "All Red Route." The All Red Route was part of the British government's policy to connect its far-flung empire by British-owned telegraph facilities to give it control over communications in times of international strife. To strengthen its control, the British government purchased Direct Cable in 1920, including the Harbour Grace station and the transatlantic cable.[87] Coincident with this, the western end of the cable was transferred from New York to Halifax.[88] This move put two transatlantic cables to Canada in British hands: the first, a German cable which the British seized during World War I, and the second, the Harbour Grace cable.

The land-line arrangement between Harbour Grace and Bay Roberts was discontinued and the Harbour Grace station again operated its transatlantic cable in competition with the others. In 1922, the company transferred the eastern end of the transatlantic cable from Ireland to Penzance.[89]

In 1929, the operation became part of the Imperial and International Communications Company, whose name was changed to Cable and Wireless Limited in 1934.[90] Shortly after Newfoundland joined Canada in 1949, Cable and Wireless's assets in Newfoundland were taken over by Canadian Overseas Telecommunication Corporation (COTC). At the same time the European terminus of the cable was transferred to Porthcurno, Cornwall.

HARBOUR GRACE FIRE

On 17 August 1944, the Harbour Grace cable station barely survived a disastrous fire that consumed a large part of the business district as well as many residences in the town. The fire started about a half mile east of the station, but high winds carried embers from the fire for miles around. Mr. F. G. Websper, along with his staff, several carpenters who were working in the building, as well as a number of soldiers guarding the building, engaged themselves in putting out fires in and around the station. The tar roof was particularly susceptible and was therefore covered with lime and other fire-retardant material. Staff member F. P. Legge lost his house and possessions in the fire and was also injured when a wall collapsed on him. In a letter to the Chairman of Cable and Wireless, Websper paid particular praise to R. E. Smith, D. M. Burke, E. J. Farnham, and C. Pike for their brave effort in saving the building from destruction.

DETERIORATION OF THE CABLE

A year earlier, in 1943, Cable and Wireless's transatlantic cable had developed problems and could no longer be used. Despite the fact that World War II was raging, attempts were made to repair the cable, but to no avail. In the meantime, Cable and Wireless continued to use the Halifax cable for telegraph business. It was not until 1951 that a serious effort was made to put the transatlantic cable back in operation. After seven years of neglect the amount of repairs required was extensive. The 8050 ton *H.M.T.S. Monarch* and the *C.S. Lady Dennison Pender* surveyed the condition of the cable and repaired the outstanding breaks. It was discovered, however, that there was an 800 mile gap that had to be completely replaced approximately 218 nautical miles from Harbour Grace and 1134 nautical miles from Porthcurno. It was also discovered that the elements had substantially deteriorated the Halifax section of the cable, and about 400 nautical miles also had to be replaced. The cost of this repair job was estimated at about $5.3 million.[91]

In the summer of 1952, the *Monarch* placed the new cable, and on 6 August reestablished communications between Harbour Grace and Porthcurno for the first time in nine years. At that time, Mr. C. H. Ryde was COTC's manager at Harbour Grace. Other staff members included D. M. Burke, C. J. Coe, G. C. R. Downer, E. J. Farnham, J. H. Jones, F. P. Legge, and B. Martin. Earlier, in 1950, employees at the cable office also included R. N. Legge and H. L. Denton, the manager of the office, who returned to England in August 1950,

By 21 August 1952, the *Monarch* had also repaired and replaced parts of the cable to Halifax, and reestablished that part of the circuit. The *Monarch* was greeted in Halifax by Premier Angus MacDonald, and D. F. Bowie, the President and General Manager of COTC, along with other dignitaries. After a long period of intermittent service, the Harbour Grace cable station was finally in full operation again, but for efficiency reasons, it was decided to transfer the station's operations to St. John's. Burke, who had joined the company around 1940, took over as acting manager, and over saw the closing of the office in 1953 when COTC moved its cable terminus to St. John's.

The staff of the Harbour Grace cable station, circa 1947. Back row: Eugene Farnham, Franklin Legge, Neil Legge, Harry Jones. Middle row: John Eades, R. V. C. Middleton (manager), Doug Burke. Front: Basil Martin (Courtesy of Mr. Basil Martin)

Commercial Cable Office, 111 Water Street, circa 1955 (Courtesy of PANL)

Thirteen

COMMERCIAL CABLE COMPANY – ST. JOHN'S

The Commercial Cable Company (Commercial Cable) was established in 1883 by John William Mackay (1811-1902), a New York mining entrepreneur, along with James Gordon Bennett (1841-1918), the owner of the *New York Herald* and several other newspapers. Bennett's newspaper empire depended on transatlantic telegraph messages for news from Europe and overseas, and he thought the rate of fifty cents a word for messages was excessive. He therefore partnered with Mackay to form an enterprise to compete with the existing transatlantic telegraph companies.[92]

In 1884, Commercial Cable contracted with Siemens for two cables between Canso, Nova Scotia and Waterville, County Kerry, Ireland. It was decided to install two cables to guard against the total loss of business in case one of them failed. The western ends of the cables were extended from Canso to Rockport, Massachusetts and Far Rockaway, Long Island, and were opened for public service on Christmas Eve in 1884. The new system was an immediate success, and Commercial Cable began to take business away from Western Union.[93] In Nova Scotia, the system was connected into Canadian Pacific Railroad Company's trans-Canada facilities, providing access to Canadian users. Service was further improved in 1885 when the eastern ends of the cables were extended from Ireland to Bristol, England and Le Havre, France. [94] The cost of this project was more than $7 million.

Commercial Cable installed a third cable from New York to England in 1894. It was laid by the *Faraday* and followed the same route as the earlier ones. In 1900, Telcon laid another cable for the company between Far Rockaway, New York, via Canso and Horta to England.[95] The same year, Commercial Cable further increased its transatlantic capacity when Siemens installed a cable between Nova Scotia and the Azores, which in 1901, was extended to New York and Waterville, Ireland. Expansion continued in 1905 when the *Colonia* installed another cable from Canso to Ireland.[96]

After Anglo-American's monopoly in Newfoundland expired, Commercial Cable decided that there would be operational advantages if it could divert its transatlantic cables from Nova Scotia to Newfoundland. In 1905, the company approached the government of Newfoundland and arranged a ten year operating agreement which allowed the company to land cables in the country. The company was also given the right to operate the government's cable across the Cabot Strait; however it was not permitted to transmit or receive telegraph messages into or out of Newfoundland. In 1909, the agreement was amended and extended to twenty-five years. The new agreement contained a provision regarding taxes which was to become a major point of contention between the company and the government. A clause in the agreement stated "In assessing any tax on the cable of the company, a cable running to the Company's station in Newfoundland and thence proceeding out again on the high seas shall be considered as one cable,..." The 1909 agreement was approved by the Governor-in-Council. Before it could be ratified by the legislature, however, there was a change in government and the agreement was repudiated, leading to a legal tax issue, discussed in the next section.

With an agreement now in place, Commercial Cables in 1909 proceeded to divert its two 1884 transatlantic cables to Cuckold's Cove, just north of St. John's. The Atlantic section of the first cable was transferred on 14 July and the western section was extended to New York on 2 August. In 1910, the company diverted the Atlantic portion of the second cable and also laid a new cable from St. John's to New York. This gave St. John's two direct telegraphic links to Ireland and New York. The work was done by the *Mackay-Bennett*, a cable ship appropriately named after Commercial Cable's founders. On a more morbid occasion, the *Mackay-Bennett* was used to retrieve bodies from the 1912 Titanic disaster.[97]

One of the company's early employees at Cuckold's Cove was Harold Brown, who was transferred there from Commercial Cable's operation in Canso. His father was also in the cable business and was superintendent at the Western Union station in Canso.

The Cuckold's Cove office remained in operation from 1909 to 1916, during which time it served as a relay station for messages between Europe and America. The rusty remains of the cables can still be seen on the beach at Cuckold's Cove.

In 1916, the company moved to a new office at 111 Water Street (a building which in recent years was Javelin House, and is now the Brother T. Murphy Centre). This building was one of the best constructed in St. John's at the time. It contained modern offices for staff, a public office, an operating room, artificial line room, test room and all the necessities for a modern telegraph operation. The building was constructed almost entirely of brick, concrete and steel to ensure against fire. The station was equipped with modern telegraph equipment, powered by strings of batteries which took up a large amount of space. Commercial power was used to light the building and charge the batteries, but the office also had its own emergency gasoline-powered generator, which provided electricity when commercial power failed. Unlike the cable companies in Heart's Content and Bay Roberts, Commercial Cable did not construct residences for its staff because sufficient accommodation was easily found in St. John's.

After the new office went into service, the Cuckold's Cove's telegraph receiving equipment was scrapped, and the old building was eventually demolished. In the early 1950s, Commercial Cable constructed another site in the area when it erected a concrete building in Quidi Vidi to house termination equipment for the overseas cables as well as equipment to amplify the telegraph signals before they were routed to Water Street.

THE TAX ISSUE

In 1905, the Newfoundland government exploited the country's favourable geographic location by introducing an annual tax of $4,000 for each transatlantic cable landed. At that time Anglo-American was the only cable company with a presence in Newfoundland. After other cable companies located in the country, a subsequent government applied this tax to all cable landings, meaning that cables transiting Newfoundland were taxed at the rate of $8,000 per year. Commercial Cable challenged the tax in court and legal battles on the government's right to impose a landing fee went on for years. Because Commercial Cable refused to pay the tax, the government gave notice in 1918 to Mr. H. D. Windeler, Commercial Cable's superintendent, that it was terminating the 1905 agreement. This action cut off the government's local cable traffic as well as all overseas traffic handled by its telegraph offices. On 1 August 1918, all government telegraph traffic was rerouted via Anglo-American.

Commercial Cable's lawyer on the tax case was Mr. William R. Howley, K.C., while the government's lawyer was Mr. Alfred B. Morine, K.C. After going to trial, Mr. Justice Kent of the Newfoundland Supreme Court ruled in favour of the government and Commercial Cable had to pay up its arrears. Commercial Cable appealed this decision, but the court's decision was upheld.

After a long period of legal disputes, Commercial Cable and the government settled their differences in 1922. As part of this settlement, the government agreed to upgrade its St. John's to Port-aux-Basques cable system and to sell to Commercial Cable its underwater cable across the Cabot Strait for $60,000. The submarine cable extended from Port-aux-Basques to Canso, where it connected into the company's Nova Scotian system. At the time of its purchase, the cable was not in operation, and Commercial Cable was obliged to effect repairs. The government also agreed to pay Commercial Cable $4,000 per year starting 23 June 1923 and to provide it with a telegraph connection between Port-aux-Basques and St. John's at no charge. An integral part of the agreement was that both parties were to carry each other's telegraph traffic on their respective telegraph systems. Commercial Cable was now satisfied, because for the first time it could transmit and receive local telegraph messages to and from Newfoundland.[98]

CODE BOOKS

Since the cost of transatlantic messages was based on the number of words, it was common for customers to send messages in code so that one word rather than many was used to represent a thought or a sentence. Code books were published containing thousands of codes for various types of telegrams. It is not known if the following telegram from the Colonial Secretary to London

CABLEGRAM

PACIFIC   ATLANTIC

NO._____TIME_____CHECK_____VIA_____ May 27th _____192 2

Send the following Cablegram
conditions printed on the back   **"VIA COMMERCIAL"**   subject to the terms and
hereof, which are agreed to.

Rurality,

London.

GINGERLY COMAWASP ROLDONES ADOSETA COMMERCIAL BUXINE FLAXSEED
SYRINGEGUN FLOURTUB BUFFOES IMITATRESS MIRECROW BUTTERFLY
SUBOPTIC BUXINE.

COLONIAL SECRETARY.

TELEPHONE - SEE OVER **FULL-RATE MESSAGE UNLESS MARKED DEFERRED**

Commercial Cables coded CABLEGRAM – 1922 (Courtesy of PANL)

Commercial Cable Company messenger boys in the mid 1950s
(Courtesy of Mrs. Ted Withers)

is coded for security reasons or to reduce the cost of the message; however, the former is more likely the case. Regardless, this Commercial Cable Cablegram is an example of what a coded message would look like.

WATER STREET BRANCH OFFICE

To handle the additional local telegraph business and to relieve pressure on its 111 Water Street location, Commercial Cable opened a branch office in St. John's. This was located at the Pope Building at 242 Water Street, at the bottom of McBride's

Commercial Cable's Office at 242 Water Street circa 1947. Note the messenger boys waiting for telegrams to be delivered. (Daily News)

Hill, presently the site of the Canada Trust building. The new office operated twenty-four hours a day and seven days a week and was designated by the company as "ND," whereas 111 Water Street was "NF." The branch office stayed in operation until the late 1940s. Immediately across Water Street, where the Scotia tower now stands, Anglo-American also had a public telegraph office.

FURTHER EXPANSION

In 1923, Commercial Cable contracted with Telcon for the installation of a duplex "Jumbo" cable between Far Rockaway (New York), Canso (Nova Scotia), Horta (Azores), and Waterville (Ireland). The cable was later extended to Weston, England and Le Havre, France[99] and was capable of a high transmission speed.

Commercial Cable diverted another two of its Canso cables (1894 and 1905) to St. John's in 1926. Both were landed at Quidi Vidi Harbour and worked in duplex mode. These cables, along with those from Cuckold's Cove, were buried to the east end of Quidi Vidi Lake. From there they were placed along the bottom of the lake to its western end, where they were trenched to the company's office at 111 Water Street.

Commercial Cable joined with Mackay Radio and Telegraph and All America Cables in 1928 to become American Cable and Radio Corporation, whose major shareholder was International Telephone and Telegraph (ITT). Commercial Cable however, continued to operate under its own name.[100] At the time, Commercial Cable's St. John's office was connected by four cables to Ireland, three to Canso and two to New York. The company had operations worldwide, with facilities in Europe, Asia, and South America

In 1941, Commercial Cable inaugurated a direct duplex printer circuit between its Water Street offices and the offices of Canadian Pacific Railway Telegraphs in Montreal and Halifax.[101] At the time, the St. John's Commercial Cable office at 111 Water Street was already directly connected with London, New York, Paris, and Canso. During World War II, all messages sent from the cable stations in Newfoundland were censored, and at the Commercial Cable office, a government official was on hand at all times to ensure that wartime security was not compromised on outgoing telegrams. The station was under constant guard and the public office section of the building was separated from the remainder by a wire screen.

At the end of World War II, the company employed about fifty people. In addition to public telegraph service, the company also provided direct private lines to major St. John's businesses such as Bowrings, Crosbies, Ayres, and Royal Stores as well as to Gander Airport. This allowed telegrams to be directly sent and received at these offices. In 1947, the employees at the St. John's station included J. MacIntyre, superintendent; A. L. Collins, Branch Office Manager; Mr. Adams, Assistant Manager; and Ira E. H. Barrett.

The 15 June 1959 St. John's *Evening Telegram* reported that the cable from Cuckold's Cove to Waterville had been ripped up by an iceberg and was out of service leaving five working cables to the mainland and one to Europe. Also reported was that at Quidi Vidi gut there were three cables in service, providing twelve circuits.

In 1960, Commercial Cable operated thirteen duplex cable circuits, providing twenty-six channels for telegraphy. Its Canadian revenues were only $397,000, a small amount compared to COTC's revenues of approximately $2.6 million. Commercial Cable realized that its old cables could not compete with the new repeater cables and decided to transfer its circuits to COTC facilities. This allowed Commercial Cable to provide service to its customers without the need for personnel in Newfoundland. Because

Mr. A. L. Collins attending equipment in the artificial line room at Commercial Cable's main building at 111 Water Street circa 1947. (Daily News)

of changes in technology and the reduction in telegraph business brought about by the increasing use of transatlantic telephone service, Commercial Cable discontinued operation at the end of 1961, when it closed down its St. John's office, affecting approximately forty employees.

QUARTER CENTURY CLUB

In 1954, Commercial Cable held a special dinner at the Newfoundland Hotel for the Edward Withers chapter of the "Quarter Century" club. This was an association of employees who had served more than twenty five years with the company. Withers had joined Commercial Cables in Canso in 1889, and was superintendent in St. John's from 1924 until his retirement in 1932. His son Edward J.(Jack) and grandson Edward (Ted) C. also worked for the company.

The Quarter Century Club members honoured at the 1954 dinner (along with their initial hiring dates) were: J. MacIntyre, July, 1901; A. Collins, August 1910; F. J. Armstrong, June 1914; G. W. Foley, November 1915; R. R. Vavasour, June 1916; H. P. Butt, November 1916; H. G. Clark, April 1919; E. J. Withers, November 1920; A. Bragg, June 1922; C. F. Collier, July 1922; J.

The members of the staff of the Commercial Cable Company at its first annual dinner of the St. John's Quarter Century Club held at the Newfoundland Hotel, 16 February 1954. Seated at the left, front to back are: Jack Adams, Harold Bradley, Ralph Vavasour, James Linegar, Frank Doyle and Gerry Foley. On the right, front to back are Charles Collier, Leo Stapleton, Art Bragg, Hugh Fardy, Ira Barrett, Frank Armstrong, J. M. MacIntyre, Jack Withers, Harvey Butt and Harry Clarke. (Daily News)

W. Linegar, July 1922; A. J. Adams, July 1922; H. J. Fardy, January 1923; L. M. Stapleton, July 1923; F. M. Ruggles, February 1925; H. S. Bradley, March 1926; F. J. Doyle, July 1926; S. J. S. Woods, December 1926; I. E. H. Barrett, July 1927.[102]

Edward J. Withers was author of the "Irene B. Mellon" radio program, a popular radio program in the 1930s. He was also Commercial Cable's last employee in St. John's after the station ceased operation in 1961, and spent the early months of 1962 decommissioning the building.

James MacIntyre had become manager of Commercial Cable's Newfoundland division in 1932. He arrived in Newfoundland after serving with the company in Glasgow, Liverpool, Canso, Havana and New York, and worked there during its first cable installation in 1909. From the early 1950s until the office closed in 1961, the superintendent was Ira E. Barrett.

Some of Commercial Cables' employees during the last year of operation were, in addition to some members of the Quarter Century Club (Ira Barrett, Harold Bradley, and Edward J. Withers), Don Clarke, James Linegar, Edgar Parke, Bruce Perry,

Commercial Cable Company bowling champions 1955-56. Left to right: Harry Clarke, Ron Stapleton, Bill Abbott, Bruce Perry, Ted Withers, Ira Barrett. (Courtesy of Mrs. Ted Withers)

Harry Rideout, William B. Trickett, William Hubert Trickett, Richard Wade, Fred White, George Whiteway, and Edward C. Withers. About a year after it closed its St. John's office, Commercial Cable also closed its office in Hazel Hill, just two miles outside Canso, Nova Scotia.[103]

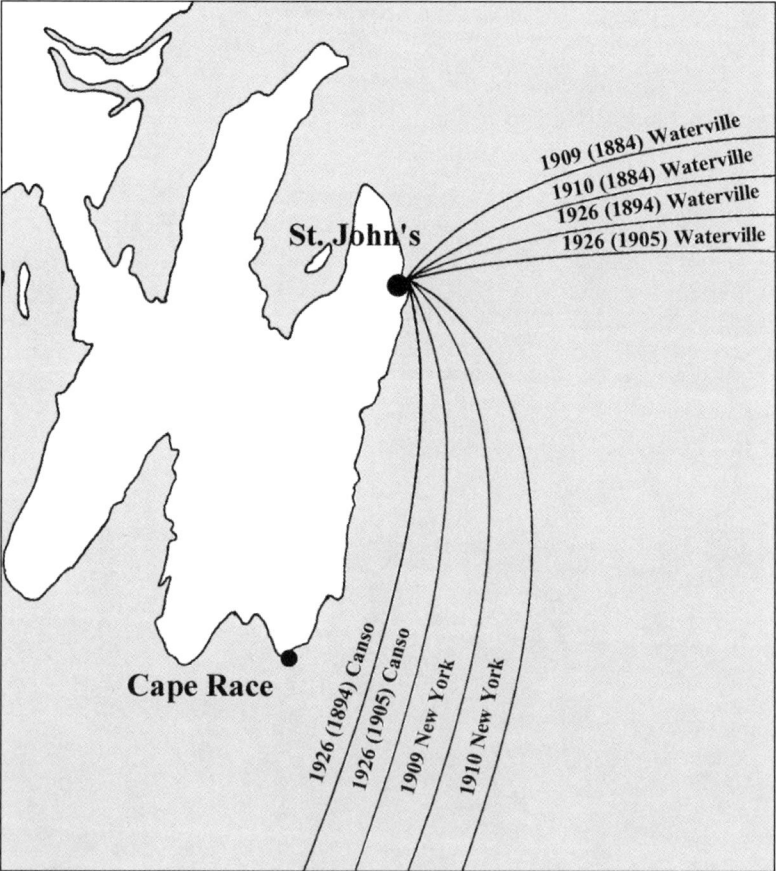

Major Commercial Cable Company cables in Newfoundland

Some of the staff of the Commercial Cable Company circa 1960. Front row: Ralph Vavasour, Jack Withers, Ira Barrett, Jack Woods, Harvey Butt. Second row: Roy Rabbitts, Frank Doyle, Charles Collier, Malcolm Vavasour. Third row: Cal Reid, Ted Withers. Fourth row: Edward Noah, Victor Parsons, Harold Bradley, Bruce Perry, William B. Trickett. Back row: Dick Wade, Leo Stapleton, Ronald Stapleton. (Courtesy of Mr. W. Hubert Trickett)

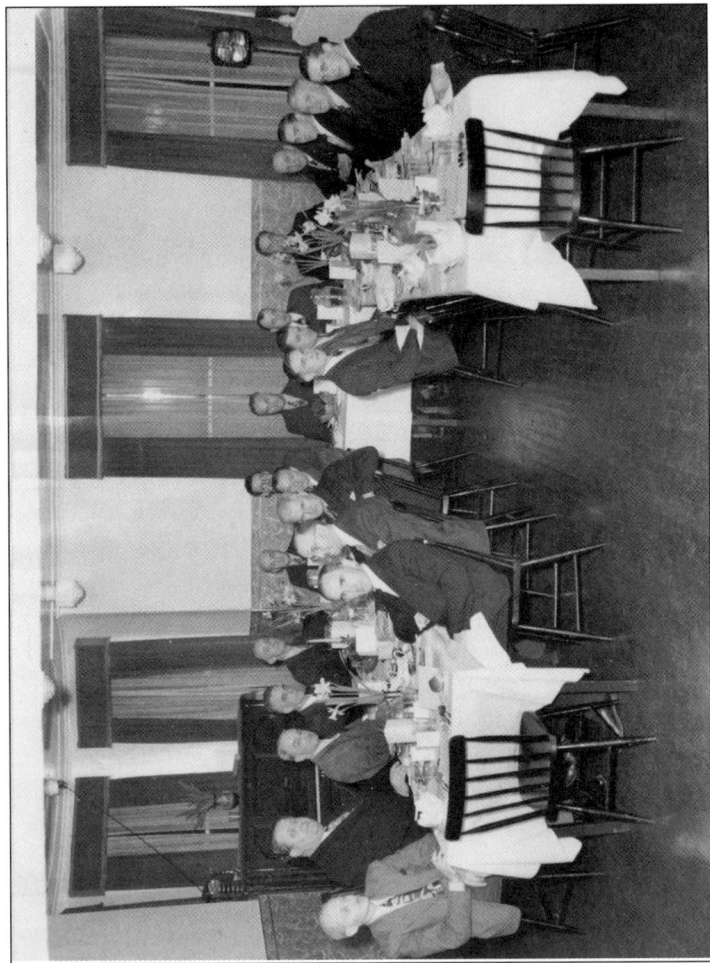

Commercial Cable Company Dinner, mid 1940s: The participants front to rear are – Left table left side: Charles Collier; Walter Chambers (entertainer), Jack Adams, Jack Woods, Fred Ruggles. Left table right side: Leo Stapleton, Roy Rabbitts, Gerry Foley, James Linnegar. Right table left side: Bruce Perry, Harold Bradley. Right table right side: Ches Atkins, Ralph Vavasour, Ted Withers, Eric Spicer. Head table left to right: Frank Armstrong, Harry Clarke, Ira Barrett, Harvey Butt, Jack Withers. (Courtesy of Mrs. Ted Withers)

Fourteen

FROM TELEGRAPH TO TELEPHONE CABLES

By 1874, the south Atlantic had also been traversed with a telegraph cable. The first telegraph connection between South America and Britain ran from Pernambuco, Brazil, via St. Vincent and Madeira to Carcavellos, Portugal.[104]

The first trans-Pacific cable was laid in 1902 from San Francisco to Hawaii, and extended to Manilla, Philippines in 1904, a distance of more than 7600 miles. In 1906, the cable was extended to Shanghai.

A cable was earlier installed from Canada, via Hawaii, to New Zealand and Australia. The project was headed by Sanford Fleming, chief engineer with Canadian Pacific. The cable was laid in 1901 from Vancouver to Fanning Island (3459 nautical miles) by the *Colonia*. The *Anglia* extended the cable to Fiji and Norfolk Island, and to Southport and Sydney, Australia as well as to Auckland, New Zealand.[105] This cable was inaugurated 1 November 1902 in a message from King Edward to Fleming. At that time it took sixty minutes to send a message from London to Australia and the rate for a message from Vancouver to Australia was fifty cents a word. Just past the turn of the century, most of the world was connected with telegraph lines, and the technology had evolved so that Cyrus Field's major feat of forty years earlier was now a commonplace event.

EARTHQUAKE

On 18 November 1929, an earthquake on the south coast of Newfoundland put most of the transatlantic cables terminating in Newfoundland out of commission, resulting in long outages and costly repairs. Seven cable ships rushed to the area to make repairs to thirteen of the twenty-one transatlantic cables that were affected. The damage was extensive: large portions of the cables were mangled and had to be replaced. The earthquake severely rearranged the ocean bottom, leaving one of the cables buried for about one hundred and thirty miles.

During the depression years of the 1930s, no transatlantic cables were installed. In the following decade, the volume of transatlantic telegraph traffic increased, especially during the war years, when Newfoundland's cable stations served as important communications links. The transatlantic telegraph cables continued to be used up to the 1960s, by which time they had become obsolete, and attention began to focus on cables for telephone service.

TELEPHONE CABLES

For almost 100 years after the first transatlantic telegraph cable, there was still no telephone cable spanning the Atlantic Ocean. Telephone communications, unlike telegraph, required a strong electrical signal, and needed to be amplified every thirty or forty miles. This meant that a transatlantic telephone cable would require fifty or sixty amplifiers (or repeaters) installed along its length. Repeaters with the high reliability requirements for a transatlantic cable lying two to three miles below the surface of the ocean were not available. Further, the introduction of radio telephone circuits across the Atlantic in 1927 reduced the urgency for telephone cables because radio technology was much cheaper than cable. Transatlantic radio transmission, however, had reliability problems and was affected by time of day and atmospheric conditions. After World War II, the necessity for reliable communications between North America and Europe was clearly established and engineering on a transatlantic telephone cable began in earnest. After years of extensive testing of various engineering designs, two different concepts emerged, one American, the other British.

The American design was the so called "flexible" repeater because the amplifiers were integrated into the cable as part of its manufacture. They were less than two inches in diameter, developed by Bell Labs and built by Western Electric. Because of their small size, the repeaters could only amplify signals in one direction, which necessitated the requirement for two cables, one to transmit and other to receive.

The British design was much larger in size and was called the "rigid" repeater. It was about ten inches in diameter and nine feet long and was engineered for use in shallow water, where a much more rugged design was required because of icebergs, fishing ships and wave action. Unlike the flexible repeater, the rigid repeater could amplify signals in both directions at the same time. This repeater was built by Standard Telephones and Cables Ltd.

Both repeaters were powered by a direct current voltage applied to the cable core from both ends of the cable. Since transistor technology was just developing, both types of repeaters used vacuum tubes. If any vacuum tube in the system failed, the whole cable would fail, so they were engineered to very high standards. The repeaters were also designed so that the location of any failed tube could be determined, to allow the cable repair ships to effect repairs.

For a transatlantic telephone, a compromise design would be used – the American flexible repeaters for the deep water portion of the cable and the British rigid repeaters for the shallow shore end. With the engineering work completed, installation of the first transatlantic telephone cable was ready to begin, and Newfoundland again played a significant role.

The CCGS John Cabot in St. John's (Courtesy of the Department of Fisheries and Oceans)

Fifteen

CANADIAN OVERSEAS TELECOM-
MUNICATION CORPORATION

The Canadian Overseas Telecommunication Corporation (COTC) was established by the Canadian government in 1950 to acquire the Canadian assets of Cable and Wireless Limited and certain assets of Canadian Marconi. COTC changed its name to Teleglobe Canada in 1975. As previously noted, the company took over Cable and Wireless's station in Harbour Grace shortly after Newfoundland's confederation with Canada in 1949.

In 1953, the 8000-ton cable ship *Monarch* diverted the Harbour Grace – Cornwall telegraph cable to Middle Cove, about seven and a half miles from St. John's. The *Monarch* laid a new section of polyethylene cable to a point about 240 miles east of St. John's, where it connected into the cable laid to Harbour Grace the previous year. From a small cable building close to the beach at Middle Cove, the cable was trenched to the new COTC office at Water Street in St. John's. The cable was buried from Middle Cove to Torbay Road, west on Elizabeth Avenue, south on Carpasian Road and Monkstown Road, and from there down the west side of Prescott Street to Water Street, and then east to the COTC building. The company considered bringing the cable into St. John's harbour, but decided that laying a cable through the narrows would be technically difficult. This was the last telegraph cable landed in Newfoundland, and upon its completion, the Harbour Grace station was closed down.

After years of research and engineering, it was finally determined that a telephone cable connecting Europe and North

America would be feasible. Installation of the first transatlantic telephone cable began on 22 June 1955 when the *Monarch* left Clarenville for Oban, Scotland paying out the "TAT-1" cable. The ship could not carry all the cable in one load and it was therefore laid in three sections. The first section was the heavy shore-end cable designed for shallow water, and extended two hundred nautical miles east of Clarenville; the second section was the deep water portion which extended 1250 nautical miles to Rockall Bank; and the third and final section extended 500 nautical miles to Oban.[106] The $42 million venture was financed 50% by American Telephone and Telegraph Company (AT&T) and its subsidiary Eastern Telephone and Telegraph Company (ET&T), 41% by the British Post Office and 9% by COTC.[107] In Clarenville, the *Monarch's* expedition created the same excitement that had surrounded the completion of the first transatlantic telegraph cable almost one hundred years earlier. A civic holiday was declared and the cable was ceremonially "christened" with water from Heart's Content harbour.

TAT-1 consisted of two 1.3 inch diameter cables manufactured by Submarine Cables Limited, the successor to the merger in 1935 of the Telegraph Maintenance and Construction Company and the cable operations of Siemens Brothers. The *Monarch* completed the second cable on 14 August 1956 and overseas telephone service began on 25 September. One cable carried telephone traffic originating in North America, while the other carried traffic originating in Europe. TAT-1 had a capacity of thirty-six voice circuits and used fifty-one repeaters to boost the volume of telephone signals. Twenty-nine circuits were used for service between London and New York, six were dedicated to London-Montreal traffic, and a single circuit was used for a number of telegraph circuits.

The TAT-1 system extended overland from Clarenville to Terrenceville on the Burin peninsula. In May 1956 it was extended by submarine cable to Sydney Mines, Nova Scotia, using fourteen repeaters. On the first full day of service, almost six hundred telephone calls were made, and by December the average number of calls was almost nine hundred per day. In addition to transatlantic telephone traffic, the cable also carried telephone calls between Newfoundland and the Canadian mainland. The TAT-1 cable remained in service until 1978.

In 1959, another transatlantic telephone facility was installed. Appropriately named TAT-2, it also used two cables, employing fifty-seven repeaters in each cable and providing thirty-six circuits.[108] This cable followed the same route as TAT-1, but terminated in Penmarch, France, rather than Scotland. The cable was owned 64.5% by AT&T, 17.7% by the French Ministry of Posts, Telegraphs and Telephones, and 17.8% by the German Federal Ministry of Posts and Telecommunications.[109] The Clarenville cable station, which terminated the TAT-1 and TAT-2 cables, was managed by the Eastern Telephone and Telegraph Company, a subsidiary of AT&T. Among the employees at the Clarenville station around 1980 were Sidney Adams, Norman Crocker, Ernest Drover (manager), George Dupont, Eugene Ploughman, Ashton Short, and Austin Tavenor. The cable station remained in service until TAT-2 was retired in 1982.

In the late 1950s, the cold war was at its height and a communications facility was required between United States Air Force bases in Greenland and NORAD headquarters in Colorado Springs, Colorado, as part of the Ballistic Missile Early Warning System (BMEWS). In 1957 COTC, in cooperation with the USAF, installed a cable between Thule, Greenland and Cape Dyer on Baffin Island. In 1961 this was extended from Cape Dyer to Hampden, White Bay and then on to Deer Lake, where it connected into microwave facilities to Colorado Springs. The cable was replaced by a direct Thule – Hampden cable in 1964. The cables were buried overland from Hampden to Deer Lake.

At Hampden, a small concrete hut was used for terminating the Atlantic cables. At Deer Lake, COTC had an office building with about fifteen employees who provided technical support. The building had thick concrete walls, and was considered "bombproof," an important consideration in light of the military significance of the station. The station also had a diesel "no-break" generating system to provide backup electric power.

COTC employees at Deer Lake in 1965 included Jerry Campbell, John Dunphy, Gerry Foley, Don Hayes, Ned Hounsell, Roger Judge, Gus Kerwin, Franklin Luther, Don Moss, Ray Rowsell, Bill Simpson, Bill Stratton, Lou Warren, Wycliffe Wellon, Alton Whelan, and John Vienneau. Most of these were reassigned to other COTC offices outside the province after the office closed.

A new telephone cable was installed between Europe and North America in 1961. The cable was named "CANTAT-1" and ran between Oban, Scotland and Hampden, White Bay. The transatlantic section was installed by the *Monarch*, and the shore end at Hampden was laid by the *Albert J. Myer*.[110] The cable was the result of a 1957 agreement between the Canadian and British governments to install a large capacity cable between the two countries, as part of a system connecting countries of the British Commonwealth. The cable was trenched from Hampden to the Deer Lake station, submerged in Deer Lake, and then again trenched to Wild Cove, near Corner Brook. From Wild Cove, the circuits were placed on COTC's 120-telephone circuit, 400-mile long submarine cable to Gross Roche, Quebec on the south shore of the St. Lawrence River, which was earlier laid by the British cable ship *Alert*. The Wild Cove building was identical to the one in Deer Lake.

The Wild Cove cable station in the early 1960s (Courtesy of Mr. Neil Legge)

Unlike the earlier TAT systems which used two cables, CANTAT-1 was the first transatlantic cable to carry telephone speech simultaneously in both directions. It was also the first transatlantic cable covered in a strong lightweight plastic, rather than the heavy metal armour previously used. It had a capacity of sixty (later increased to eighty) voice circuits and was the first

Staff of COTC's Wild Cove station circa 1964. Front, left to right: Arnold Bennett, Gerard Peddle, Lloyd Way, John Neilson, Ross Lee. Back, left to right: Neil Legge, Harvey Hoffe, Jack Reeves, Bill Stratton, Eric Pennell, Dave Stewart, Hiram Tiller, Malcolm Bayne. (Courtesy of Mr. Neil Legge)

transoceanic cable to employ rigid repeaters. At the time, CAN-TAT-1 was the most technologically advanced cable in the world. It was laid in three operations from Oban, and was armoured on the continental shelf portion to protect it from iceberg and fishing activity. The approximately 2000 nautical miles of cable required 86 repeaters.[111] The new system was officially placed in service on 19 November 1961 in a telephone call between Queen Elizabeth II in England and Prime Minister John Diefenbaker in Ottawa.

On 1 January 1963, another cable went into service when ICECAN was inaugurated between Canada, Greenland and Iceland, where it connected into the SCOTICE cable to Great Britain.[112] The cable landed at Hampden, White Bay and extended underground via the Deer Lake station to Wild Cove. The cable provided twenty-four voice circuits primarily for the International Civil Aviation Organization, as well as for telephone circuits to Britain and Europe.

COTC established a storage depot in St. John's to provide maintenance material for the repair of North Atlantic cables. The facility included eight storage tanks which held up to sixteen hundred miles of various submarine repair cable. Also stationed in St. John's was the *John Cabot*, an ice-breaker/cable ship, which was owned by the Canadian Coast Guard but chartered to COTC under a long term arrangement.

In 1974, CANTAT-2 was installed between Widemouth, England and Beaver Harbour, Nova Scotia. This cable used new coaxial technology and was 1.47 inches in diameter, capable of carrying 1840 voice circuits. This vastly increased circuit capacity rendered the other transatlantic telephone cables obsolete. TAT-1 was retired in 1978, TAT-2 in 1982, and CANTAT-1 in 1985. The closure of these cables made COTC's Newfoundland stations in St. John's, Deer Lake, and Corner Brook redundant and they were all decommissioned.

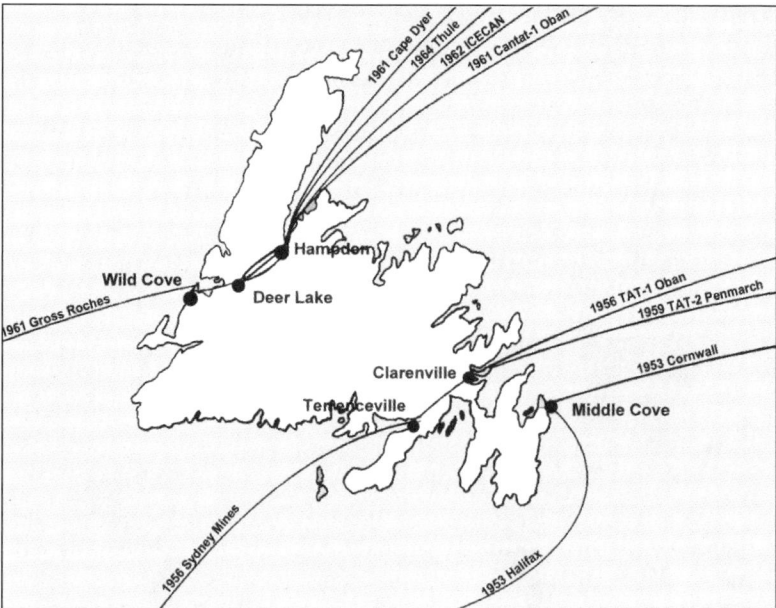

Major cable landings in Newfoundland in the 1950s and 1960s operated by COTC

Sixteen

END OF AN ERA

There are presently only two long distance submarine cables landing in Newfoundland. Newfoundland Telephone and Maritime Telegraph and Telephone installed an optical fibre cable between Cape Ray, Newfoundland and Sydney Mines, Nova Scotia in 1991. This cable was 180 kilometres in length, the longest stretch of non-repeatered fibre optic cable in the world.[113] A second Cabot Strait optical fibre cable was installed in 1996 between Dingwall, Cape Breton Island and Codroy, on the west coast of Newfoundland. These cables carry video, high speed data and telephone circuits and are technologically light years ahead of the old telegraph cables, all of which have long since been decommissioned.

In this short account, it has only been possible to cover the major highlights of the history of submarine telegraph cables in the province. Over the years, many telegraph and telephone cables were landed, but by the late 1990s, with the exception of the optical fibre cables across the Cabot Strait, all had been abandoned. The old copper core telegraph and telephone cables had fallen victim to the changing forces of technology. Microwave radio and optical fibres provided state-of-the-art communications between Newfoundland and the Canadian mainland. With modern satellite and optical-fibre systems, there was no longer any need to route intercontinental facilities via the province. The long history of transatlantic cables in Newfoundland had come to an end.

From Gisborne's first visit to Newfoundland over one hundred and fifty years ago, to the present, approximately fifty submarine telegraph cables were landed in the province. The towns of Heart's Content, Bay Roberts, Harbour Grace, St. John's, Clarenville, Deer Lake, and Wild Cove have played significant roles in the history of communications in Newfoundland and indeed the world. Newfoundland's involvement in the first transatlantic telegraph and telephone cables are significant events that the province should be proud of. It is hoped that this short history has captured some of the more glorious moments.

ENDNOTES

1. Collins, Robert, A Voice from Afar, McGraw-Hill Ryerson, Toronto, 1977, p. 20.
2. Clarke, Arthur C., Voice Across the Sea, Harper & Brothers, New York, 1958, p. 6.
3. Staiti, Paul J., Samuel F. B. Morse, Cambridge University Press, Cambridge, 1989, p. 15.
4. Collins, p. 24.
5. Oslin, George P., The Story of Telecommunications, Mercer University Press, Macon, Georgia, 1992, p. 35.
6. Oslin, p. 158.
7. Collins, p. 25.
8. Oslin, p. 159.
9. Collins, p. 31.
10. Oslin, p. 159.
11. Prowse, D. W., A History of Newfoundland, Mika Studio, Belleville, Ontario, 1972, p. 635.
12. Garratt, G. R. M., One Hundred Years of Submarine Cables, His Majesty's Stationery Office, London, 1950, p. 9.
13. PANL GN 2/5, File #22.
14. Oslin, p. 164.
15. Russell, W. H., The Atlantic Telegraph, David & Charles (publishers) Limited, Newton Abbot, 1865, p. 9.
16. Oslin, p. 165.
17. Russell, p. 10.
18. Daily News, 26 July 1916.
19. Field, Henry M., The Story of the Atlantic Telegraph, Gay and Bird, London, 1893, p. 22.
20. Lawford, G. L., and Nicholson, L. R., The Telcon Story, The Telegraph Construction and Maintenance Co. Ltd., London, 1950
21. Oslin, p. 167.
22. Oslin, p. 167.
23. Haigh, K. R., Cableships and Submarine Cables, Adlard Coles Ltd., London, 1968, p. 32.
24. Haigh, p. 38.
25. Charles Bright's log, Cable and Wireless Company, London, England
26. Oslin, p. 169.
27. Oslin, p. 169.
28. Field, p. 200.
29. Charles Bright's log.

30. Field, p. 199.
31. Charles Bright's log.
32. Charles Bright's log.
33. Oslin, p. 172.
34. Garratt, p. 7-8.
35. Oslin, p. 158.
36. Proceedings of the Royal geographic Society of Great Britain, January
    28 and February 11, 1861.
37. Collins, p. 37-45.
38. Oslin, p. 172.
39. Field, p. 232.
40. Field, p. 247.
41. Dugan, James, The Great Iron Ship, Harper & Brothers, New York,
    1953, p. 1-2.
42. Oslin, p. 174.
43. Oslin, p.174.
44. Bright, Charles, The Story of the Atlantic Cable, D. Appleton and
    Company, New York, 1903, p. 189.
45. Oslin, p. 174.
46. Colonial Commerce, 31 March 1918.
47. PANL MG 826.
48. Colonial Commerce, March 31, 1918.
49. PANL MG 826.
50. Field, p. 343.
51. Oslin, p. 176.
52. Encyclopedia of Newfoundland and Labrador, Volume Two, page 893.
53. Oslin, p. 177.
54. Oslin, p. 178.
55. Oslin, p. 178.
56. Oslin, p. 239, Haigh, p. 317.
57. Oslin, p. 239, Haigh, p. 321.
58. Haigh, p. 321.
59. Haigh, p. 330.
60. PANL MG570, Series A, Reel #1.
61. PANL MG570 Box 1, Anglo Diary 1870.
62. Haigh, p. 41.
63. Haigh, p. 246.
64. Rowe, Melvin, I Have Touched the Greatest Ship, Town Crier
    Publishing Co., 1976.
65. Haigh, p. 42, p. 246.
66. Haigh, p. 246.
67. McAlpine's 1894-95 Directory.
68. Prowse.
69. Haigh, p. 252.
70. Oslin, p. 182.
71. PANL MG 846.
72. Oslin, p. 182.
73. Haigh, p. 251.

74. Haigh, p. 251, Oslin p. 182.

75. Oslin, p. 253, Haigh, p. 42.

76. Evening Telegram, 12 August, 1992.

77. Haigh, p. 42.

78. PANL MG570, Box #1.

79. Haigh, p. 252, Coggenshall, p. 219.

80. Oslin, p. 291.

81. Haigh, p. 252.

82. Hambling, Jack, The Second Time Around, Harry Cuff Publications Ltd., St. John's, 1992, p. 15.

83. Barty-King, Hugh, Girdle Round the Earth, William Heinman Limited, London, 1979, p. 50.

84. Haigh, p.66.

85. Barty-King, p. 57.

86. Prowse, p. 497.

87. Haigh, p. 152, 251.

88. Barty-King, p. 173.

89. Haigh, p. 202.

90. Haigh, p. 152.

91. Daily News, 7 August 1952.

92. Haigh, p. 257.

93. Haigh, p. 257, Oslin, p. 239.

94. Oslin, p. 239.

95. Haigh, p. 257.

96. Oslin, p. 239.

97. Haigh, p. 258.

98. Evening Telegram, 15 May, 1947.

99. Oslin, p. 290, Haigh, p. 42.

100. Haigh, p. 258.

101. Daily News, 20 September, 1941.

102. Daily News, 17 February 1954.

103. Financial Post, October 7, 1961.

104. Haigh, p. 136.

105. Oslin, p. 253.

106. Haigh, p. 203.

107. Haigh, p. 276.

108. Barnes, p. 23.

109. Haigh, p. 277.

110. Haigh, p. 155.

111. Haigh, p. 155.

112. Haigh, p. 237, 353, 406.

113. Newfoundland Telephone Company Limited Annual Report 1991.

BIBLIOGRAPHY

Barnes, C. C., Submarine Telecommunication and Power Cables, Peter Peregrinus Ltd., Herts, England, 1977

Barty-King, Hugh, Girdle Round the Earth, William Heinmann Limited, London, 1979

Bright, Charles, The Story of the Atlantic Cable, D. Appleton and Company, New York, 1903

Clarke, Arthur C., Voice Across the Sea, Harper & Brothers, New York, 1958

Clayton, H., Atlantic Bridgehead, Granstone Press, 1968

Collins, Robert, A Voice from Afar, McGraw-Hill Ryerson, Toronto, 1977

Dugan, James, The Great Iron Ship, Harper & Brothers, New York, 1953

Dibner, B., The Atlantic Cable, Ginn and Company, Toronto, 1959

Field, Henry M., The Story of the Atlantic Telegraph, Gay and Bird, London, 1893

Finn, Bernard S., Development of Submarine Cable Communications, Arno Press, New York, 1980

Garnham, Captain S. A., and Hadfield, Robert L., The Submarine Cable, Sampson Low, Marston & Co. Ltd., London, 1934

Garratt, G. R. M., One Hundred Years of Submarine Cables, His Majesty's Stationery Office, London, 1950

Haigh, K. R., Cableships and Submarine Cables, Adlard Coles Ltd., London, 1968

Hambling, Jack, The Second Time Around, Harry Cuff Publications Ltd., St. John's, 1992

McCarthy, M., Galgay, F., OKeefe, J., The Voice of Generations, Newfoundland Telephone, St. John's, 1994

Lawford, G. L., and Nicholson, L. R., The Telcon Story, The Telegraph Construction and Maintenance Co. Ltd., London, 1950

Newfoundland Historic Trust, Ten Historic Towns, Valhalla Press, Canada, 1978

Ogle, Ed, Long Distance Please, Collins Publishers, Don Mills, 1979

Prowse, D. W., A History of Newfoundland, Mika Studio, Belleville, Ontario, 1972

Rowe, Melvin, Heart's Content Pioneer in World Communications, Pamphlet, 1972

Rowe, Melvin, I Have Touched the Greatest Ship, Town Crier Publishing Co., 1976

Russell, W. H., The Atlantic Telegraph, David & Charles (publishers) Limited, Newton Abbot, 1865

Smallwood, Joseph R., Books of Newfoundland, Volumes 1,2,3,4,5 and 6, Newfoundland Book Publishers Limited, St. John's, 1975, 1979

Smallwood, Joseph R., Encyclopedia of Newfoundland, Newfoundland Book Publishers (1967) Limited, St. John's, 1981, 1984, 1991, 1993

Staiti, Paul J., Samuel F. B. Morse, Cambridge University Press, Cambridge, 1989

APPENDIX – MAJOR SUBMARINE CABLE LANDINGS REFERENCED

YEAR	CABLE ROUTING	COMPANY	INSTALLED AND/OR MANUFACTURED BY:	SHIP	COMMENTS
1856	Cape Ray - Cape Breton Island	New York Telegraph	Kuper & Company	*Propontis*	First cable linking Newfoundland with mainland
1858	Valentia - Heart's Content	Atlantic Telegraph	Glass Elliot & Company	*Niagara, Agamemnon*	The first transatlantic cable, however worked only three weeks
1865	Valentia - Heart's Content	Atlantic Telegraph	Telcon	*Great Eastern*	The 1865 attempt did not succeed. The cable was recovered and put in service in 1866. It went out of service in 1877
1866	Valentia - Heart's Content	Anglo-American	Telcon	*Great Eastern*	This was the first commercially successful transatlantic cable. It failed in 1872
1866	CapeRay-CapeNorth,N.S.	New York Telegraph	Telcon	*Medway*	Completed after the 1866 transatlantic cable
1867	Placentia - St. Pierre - Sydney, N.S.	New York Telegraph		*Chiltern*	Built as backup to the Cape Ray overland line
1869	Brest, France - St. Pierre	Société du Cable Transatlantique Française	Telcon	*Great Eastern*	SCTF merged with Anglo-American in 1873. Cable speed was 10.5 words per minute
1869	St. Pierre - Duxbury, Massachussetts	Société du Cable Transatlantique Française		*Chiltern, William Cory, Scanderia*	Cable from Brest to Duxbury was 2685 nautical miles.

YEAR	CABLE ROUTING	COMPANY	INSTALLED AND/OR MANUFACTURED BY:	SHIP	COMMENTS
1873	Heart's Content - Rantem - Island Cove - Placentia	Anglo-American		Robert Lowe	The Robert Lowe sank enroute to Placentia in November 1873
1873	Valentia - Heart's Content	Anglo-American	Telcon	Great Eastern	Largely used for stock market reports
1873	Placentia - St. Pierre	Anglo-American	Telcon	Vanessa	
1873	Island Cove - Cape Breton Island, N.S.	Anglo - American	Telcon	Vanessa	
1874	Ballinskelligs - Torbay, N.S.	Direct Cable	Siemens	Faraday	Transferred to Harbour Grace in 1910.
1874	Valentia - Heart's Content	Anglo - American	Telcon	Great Eastern	Cable was laid from west to east. Last transatlantic cable laid by the Great Eastern
1879	Brest - St. Pierre- Cape Breton- Cape Cod - Rye Beach, U.S.	Compagnie Française du Télégraphe de Paris à New York	Siemens	Faraday	"P. Q. Company"
1880	Placentia - St. Pierre - North Sydney	Anglo-American		Seine, Kangaroo	Three-core cable
1880	Valentia - Heart's Content	Anglo-American	Telcon	Seine	Practically a rebuild of the 1866 line. Went out of operation in 1949

YEAR	CABLE ROUTING	COMPANY	INSTALLED AND/OR MANUFACTURED BY:	SHIP	COMMENTS
1881	Sennen Cove, Cornwall - Canso, Nova Scotia	American Telegraph	Siemens	*Faraday*	Diverted to Bay Roberts in 1913
1882	Sennen Cove, Cornwall - Canso	American Telegraph	Siemens	*Faraday*	Diverted to Bay Roberts in 1915
1884	Waterville - Canso	Commercial Cable	Siemens	*Faraday*	Diverted to Cuckold's Cove in 1909
1884	Waterville - Canso	Commercial Cable	Siemens	*Faraday*	Diverted to Cuckold's Cove in 1910
1890	Island Cove -St. Pierre	Anglo-American			
1893	Island Cove - North Sydney	Anglo-American			
1894	Valentia - Heart's Content	Anglo-American	Telcon	*Scotia*	Last transatlantic cable to Heart's Content
1894	U.K. - Azores - Canso	Commercial Cable	Siemens	*Faraday*	Diverted to St. John's in 1926
1898	Brest, France - Cape Cod, U.S.	Compagnie Française des Cables Télégraphiques	Silvertown Company	*François Arago*	At 3174 nautical miles, this was the longest continuous underwater cable.
1900	Borkum, Germany - Vigo - Azores - New York	German Atlantic Telegraph Company	Telcon	*Anglia, Briannia*	This was the first German transatlantic cable

YEAR	CABLE ROUTING	COMPANY	INSTALLED AND/OR MANUFACTURED BY:	SHIP	COMMENTS
1900	Nova Scotia - Azores	Commercial Cable	Siemens	*Faraday*	
1901	Azores - Waterville, Ireland	Commercial Cable	Siemens	*Faraday*	
1903	Borkum - Virgo - Azores	German Submarine Telegraph Company	Norddeutsche	*Stephan*	German Submarine Telegraph was later absorbed by German Atlantic Telegraph
1904	New York - Azores	German Submarine Telegraph Company	Norddeutsche	*Stephan*	German Submarine Telegraph was later absorbed by German Atlantic Telegraph
1905	Waterville, Ireland - Canso	Commercial Cable	Telcon	*Colonia*	Duplex. Diverted to St. John's in 1926
1909	St. John's - New York	Commercial Cable			Installed after the Canso cable was diverted
1910	St. John's - New York	Commercial Cable			Installed after the Canso cable was diverted
1910	Sennen Cove, Penzance - Bay Roberts - Coney Island	Western Union	Telcon	*Colonia*	Western Union's first transatlantic cable
1913	Bay Roberts - Colinet - North Sydney	Western Union			Landline between Bay Roberts and Colinet
1920	Placentia - St. Pierre	Western Union			

YEAR	CABLE ROUTING	COMPANY	INSTALLED AND/OR MANUFACTURED BY:	SHIP	COMMENTS
1921	Island Cove - North Sydney	Western Union			Extended to Canso in 1922
1923	Far Rockaway, New York - Canso - Horta - Waterville	Commercial Cable	Siemens and Telcon	*Faraday*, *Colonia*	"Jumbo cable"
1924	New York - Azores	Western Union	Telcon	*Robert C. Clowry*	Loaded cable, 1500 letters per minute
1926	Penzance, U.K. - Bay Roberts - Hammell, U.S.	Western Union	Telcon	*Colonia*	Loaded cable. Simplex
1928	Horta, Azores - Bay Roberts	Western Union	Telcon	*Colonia*	Loaded cable. High speed duplex
1953	Cornwall - St. John's	COTC	Cable and Wireless	*Monarch*	Major rebuild of 1874 cable. Diverted from Harbour Grace to St. John's via Middle Cove in 1953
1956	Oban, Scotland - Clarenville	COTC, AT&T, BPO	Submarine Cables Ltd.	*Monarch*	TAT-1. First transatlantic telephone cable. Retired in 1978
1957	Cape Dyer - Thule, Greenland	COTC and USAF	Simplex Wire and Cable		Retired in 1964
1959	Penmarch, France - Clarenville - Sydney	AT&T, French Post, German Post	Submarine Cables, Simplex Wire and Cable, Câbles de Lyon, Norddeutsche Seekabelwerke	*Monarch*, *Ocean Layer*, *Ampere*	TAT-2. Retired in 1982

YEAR	CABLE ROUTING	COMPANY	INSTALLED AND/OR MANUFACTURED BY:	SHIP	COMMENTS
1961	Oban, Scotland -Hampden, White Bay	COTC et al	Submarine Cables Limited	*Monarch*	CANTAT - 1 Initially carried 60 voice circuits. Retired in 1986
1961	Wild Cove - Gross Roches, Quebec	COTC	Submarine Cables Limited	*Alert, Hausund*	120 circuits
1961	Cape Dyer - Hampden	COTC , USAF	Simplex Wire and Cable		Retired in 1964
1963	Great Britain - Iceland - Greenland - Hampden	COTC , Great Northern	Norddeutsche	*Neptun*	ICECAN
1964	Thule - Hampden - Deer Lake	COTC, USAF			Retired in 1976
1974	Widemouth, England - Beaver Harbour, N. S.	COTC , BPO		*John Cabot*	CANTAT - 2, Coaxal cable, 1840 circuits. Retired in 1982
1991	Cape Ray - Sydney Mines, N.S.	Newfoundland Tel, Maritime Tel and Tel	BT Marine	*Discovery*	At the time was the world's longest non-repeatered optical fibre cable. Cable extended to St. John's
1996	Codroy, Newfoundland - Dingwall, N.S.	Newfoundland Tel, Maritime Tel and Tel	Teleglobe Marine	*John Cabot*	Cable extended to St. John's

INDEX

Adger, 23, 24
Agamemnon, 34-35, 37, 39-43, 45, 55, 57
Albany, 62, 67
Alert, 136
All America Cables, 122
All Red Route, 110, 111
American Telegraph, 97
American Telephone and Telegraph Company (AT&T), 134, 135
Amethyst, 57
Amherst, 8
Anderson, Captain James, 57
Anglia, 129
Anglo-American Telegraph Company (Anglo-American), 61-69, 75-96, 98,
       107-110, 116-118
Archibald, Edward M., 21
Archibald, Samuel G., 10, 11
Arctic, 33
Armstrong, Samuel, 49
artificial line, 94, 101, 117, 123
Associated Press, 11, 12, 43, 47
Atlantic Telegraph Company (Atlantic Telegraph), 30, 31, 33, 39, 42-46,
       53, 55-57, 61, 64, 75
Atlantic Telegraph Company, 29, 30, 72, 73
Azores, 49, 72, 100, 107, 116
Baffin Island, 135
Bailey, Samuel, 77-79, 82, 83
Ballinskelligs, 107, 108
Bay Roberts, 11, 87, 93, 97-106, 111, 117, 140
Beaver Harbour, 138
Bennett, James Gordon, 115
Berryman, Lieutenant O. H., 28, 33
Bloodhound, 64, 67
BMEWS, 135
Borkum, 72
Bowie, D. F., 112
Brassey, Thomas, 54
Brest, 71, 72, 79
Brett, John and Jacob, 15-17, 27, 29, 30, 49
Bright, Charles, 30, 32, 44, 57
Brigus, 13, 16, 77
Britannia, 85
British Columbia, 51, 52
British North American Electric Telegraph Association, 7
British Post Office, 72, 134

Brown, Harold, 117
Brown, Sir William, 30,
Brunel, Isambard Kingdom, 55
Bryant, Sarah L., 24, 25
Buchanan, President James, 43-45
Buffalo, 5, 97
Bull Arm (Sunnyside), 35, 38, 42, 43, 55
Bulldog, 51
Cable and Wireless, 73, 99, 111, 112, 133
Cable Avenue, 98, 101, 102
Cabot Strait, 9, 14, 19, 23, 25, 26, 63, 65, 67, 69, 77, 116, 118, 139
Calais, 6, 8
Canadian Marconi, 133
Canadian Overseas Telecommunication Corporation (COTC), 111, 112,
        123, 133-138
Canadian Pacific, 115, 122, 129
Canning, Samuel, 24, 57, 59, 60, 68, 70
Canso, 79, 86, 97, 98, 115-117, 122, 124-126
CANTAT-1, 136, 137
CANTAT-2, 138
Cape Breton, 9, 10,12, 14, 21, 24, 26, 67, 69, 72, 81, 139
Cape Cod, 72
Cape Dyer, 135
Cape North, 10, 12, 14
Cape Race, 11, 47, 48
Cape Ray, 9, 10, 12-16, 19, 23, 24, 26, 77, 81, 82, 139
Cape Tormentine, 16
Carbonear, 11-14, 16, 81
Chatterton, John, 55
Chiltern, 71
Clarenville, 134, 135, 140
code books, 118-121
Codroy, 139
Colinet, 98
Collett, 65, 77
Collins, Perry, 51, 52
Collins Overland Telegraph, 51, 52
Colonia, 72, 98, 99, 100, 109, 116, 129
Commercial Cable Company (Commercial Cable), 71, 87, 98, 110, 115-128
Compagnie Française des Cables Télégraphiques, 72
Compagnie Française des Télégraphes Sous-marins, 72
Compagnie Française du Télégraphe de Paris á New York, 72
Coney Island, 72
Cooke, William, 3
Cooper, Peter, 18, 20, 21
Corner Brook, 136, 138

Cornwall, 97, 111, 133
Craig, D. H., 26
Crampton, T. R., 16
Cuckold's Cove, 116, 123
Cunard Line, 45, 57
Cyclops, 33, 34
Darrow, Lawson R., 6
Dauntless, 64
Dayman, Joseph, 33
de Sauty, C. V., 57, 58
Deane, John C., 65, 66
Deer Lake, 135-138
Direct United States Cable Company (Direct Cable), 71, 87, 98, 107-113
double key, 93
duplex, 85, 94, 100, 122, 123
Duxbury, 71
earthquake, 130
Eastern Telephone and Telegraph Company, 135
Eddy, James, 26
Edinburgh, 80
Ellen Gisborne, 16
English and Irish Magnetic Company, 31
English Channel Submarine Telegraph Company, 15
Europa, 45
Everett, William, 39, 43
Far Rockaway, 115, 116, 122
Faraday, 97, 107, 108, 115
Fayal, 72
Female employees, 92
Field, Cyrus, 5, 10, 17-69
Fleming, Sanford, 129
Fox, 51
François Arago, 72
French Ministry of Posts, Telegraphs and Telephones, 135
galvanometer, 64, 93, 94
Gamecock, 83
General Oceanic and Subterranean Electric printing Telegraphic Company,
     49
German Atlantic Telegraph Company, 72
German Federal Ministry of Posts and Telecommunications, 135
Gisborne, Frederick, 6-17, 19, 21, 22, 27, 139
Glass Elliott & Company, 23, 33, 39, 54
Glass, R. A., 55
Gooch, Daniel, 56-58, 61, 65
Goodwin, Captain, 25
Gould, Jay, 97

Great Eastern, 34, 55-68, 71, 80, 83
Greenland, 50, 135, 137
Gross Roche, 136
Gutta Percha Company, 15, 33, 54
gutta-percha, 7, 15, 16, 31, 33, 49
Halifax, 2, 6, 8-10, 12, 14, 21, 23, 47, 72, 79, 108, 110, 112, 122
Halpin, Captain Robert, 71, 80, 83
Hamilton, Governor Ker Baillie, 21
Hammel, 98, 100
Hampden, 135, 136
Hanrahan, Edmund, 11
Harbour Grace, 11, 13, 64, 77, 87, 93, 99, 106-113, 133, 140
Hazel Hill, 126
Heart's Content, 62-72, 74-96, 140
Hibernia, 80
Hill, Governor Stephen John, 80
Holbrook, Darius B., 15
Horta, 100, 116, 122
Hunt, Wilson G., 18, 20
ICECAN, 137
Imperial and International Communications Company, 111
International Civil Aviation Organization, 137
Ireland, 9, 10, 15, 16, 22, 28, 29, 31, 33-35, 40, 42, 43, 57, 58, 66, 77, 83, 85, 107, 111, 115, 116, 122
Iris, 57
Island Cove, 81, 86
Johnson, President Andrew, 65, 66
Jumbo cable, 122
Kangaroo, 83, 86
Kuper & Co., 23
Labouchere, Henry, 109, 110
Lady Dennison Pender, 112
Lampson, C. M., 30
Le Havre, 115, 122
Leopard, 34
Leviathan, 56
Lilly, 66
Little, Philip F., 10
loaded cables, 100
Loughrigg Holme, 85
Lundy, 64, 65, 77
MacIntyre, James, 123-125
Mackay, Alexander M., 26, 67, 77
Mackay, John W., 115
Mackay Radio and Telegraph, 122
Mackay-Bennett, 116

Magnetic Telegraph Company, 5
Maritime Tel & Tel, 139
Maury, Lieutenant Matthew F., 28, 29
Mechanics Institute, 13
Medway, 62, 63, 67-69
Middle Cove, 133, 134
Monarch, 112, 133, 134, 136
Montreal Telegraph Company, 5
Morse, 3-6, 18, 23, 24, 27-29, 31, 32, 49
Mullock, Bishop J. T., 9, 10, 27, 49
Murphy, John, 48
Musgrave, Governor, 66
New Brunswick, 6, 8, 14, 16, 21, 24
New York, Newfoundland and London Telegraph (New York Telegraph),
       20-22, 26, 67, 75, 77, 79, 85, 109
Newall & Company, 16, 33
Newfoundland Constabulary, 87, 102
Newfoundland Electric Telegraph Company, 14-17, 20, 21
Newfoundland government, 10, 14, 25, 30, 109, 110, 117, 118
Newfoundland Telephone, 139
Niagara, 34-37, 39, 40, 42-44, 55
Niger, 62
NORAD, 135
North Sydney, 86, 98
Nova Scotia, 2, 8, 12, 14, 16, 22, 25, 26, 29, 46, 47, 49, 67, 79, 97, 107,
       108, 115, 116, 126, 134, 138
Nova Scotia Telegraph Company, 8, 26
Oban, 134, 136, 137
Oersted, 2
optical fibres, 139
P.Q. Company, 72
Pender, John, 30, 54, 107, 108
Penmarch, 135
Penn & Company, 57
Penzance, 72, 97, 98, 100, 111
Pitts, J. & W., 26
Porcupine, 42, 43
Porthcurno, 111, 112
preemption, 109, 110
Prince Edward Island, 14, 16, 21, 22
Prince of Wales, 57
Propontis, 25
Quarter Century Club, 124, 125
Queen Victoria, 44, 45, 65, 66
Quidi Vidi, 44, 117, 122, 123

Raccoon, 62

Rantem, 81, 86

repeater, 93, 94, 101, 130, 131, 134, 137

Reuter, Paul Julius, 71

Ridley Hall, 109

Roberts, Marshall, 18, 20, 21

Ronald, Sir Francis, 2

Russia, 51, 52

Rye Beach, 107

Saunders and Howell, 81

Scotia, 85

SCOTICE, 137

Seine, 83, 86

semaphore, 1, 2

Shaffner, Colonel Taliaferro, 50, 51

Siemens Brothers, 72, 97, 107, 115, 116, 134

simplex, 100

Société du Cable Translantique Française (SCTF), 71, 72, 79

Société Générale des Téléphones, 72

Sommerring, 2

Southcott, J. & J., 69, 80

Sphinx, 58, 59

St. John's, 7-17, 19, 22-26, 44, 46, 66, 67, 69, 93, 116, 117, 121-126 133

St. John's and Carbonear Electric Telegraph Company, 12-14

St. Pierre, 71, 72, 79, 85, 86, 107

Stephan, 72

Submarine Cables Limited, 134

Sunnyside (Bull Arm), 35, 38, 42, 43, 55

Sydney, 14, 77, 81, 85

Sydney Mines, 134, 139

TAT-1, 134, 135, 138

TAT-2, 135, 138

tax issue, 117, 118

Taylor, Moses, 20, 21

Tebbets, Horace, 15, 17

Technical Committee, 53-55

Telegraph Construction and Maintenance Company (Telcon), 54-57, 61, 62,
        65, 85, 98, 115, 122

Teleconia, 98, 99

Teleglobe Canada, 133

telephone cables, 129, 130, 131,133, 134, 136, 138

Terrenceville, 134

Terrible, 58, 59, 62, 67, 69

Thompson, William (Lord Kelvin), 29, 30, 57, 64

Thule, 135

Tidmarsh, Captain, 82, 83

Tor Bay, 107, 108
Toronto, Hamilton & Niagara Electro-Magnetic Telegraph Company, 5
Tranfield, C. H., 87, 93
Trepassey, 11, 12, 15
Valentia, 35-37, 39, 40, 42, 43, 45, 55, 58, 59, 62, 68, 77, 81, 82
Variety Hall, 80
Varley, 54, 57
Victoria, 21, 23
Vigo, 48
Waddell, John, 77, 78
Waterville, 115, 116, 122, 123
Websper, F. G., 111
Weedon, Ezra, 65, 77-79
Western Union, 51, 52, 71, 78, 85, 91-93, 95, 96, 98-100, 108, 113, 115
Wheatstone, Charles, 3, 54, 58, 94
White, Chandler, 18, 20-23, 60
Whitehouse, Dr. Edward , 30, 32, 46
Widemouth, 138
Wild Cove, 136, 137
William Cory, 62
Windeler, H. D., 118
Withers, Edward, 124, 125
Wollaston, C. J., 15
Wood, Orrin, 6, 7
Woolwich, 107
World War I, 72, 87, 99, 101, 111
World War II, 87, 102, 112, 122, 130